VLSI Planarization

Mathematics and Its Applications

Managing Editor:

M. HAZEWINKEL

Centre for Mathematics and Computer Science, Amsterdam, The Netherlands

Volume 399

VLSI Planarization

Methods, Models, Implementation

by

V. Feinberg

SILVACO International,
Santa Clara, California, U.S.A.

A. Levin

Intel Corporation,
Santa Clara, California, U.S.A.

and

E. Rabinovich

Gorizont,
Minsk, Belarus

SPRINGER SCIENCE+BUSINESS MEDIA, B.V.

A C.I.P. Catalogue record for this book is available from the Library of Congress.

ISBN 978-94-010-6421-7 ISBN 978-94-011-5740-7 (eBook)
DOI 10.1007/978-94-011-5740-7

Printed on acid-free paper

Table of Contents

Introduction

At the beginning we would like to introduce a refinement. The term 'VLSI planarization' means planarization of a circuit of VLSI, i.e. the embedding of a VLSI circuit in the plane by different criteria such as the minimum number of connectors, the minimum total length of connectors, the minimum number of over-the-element routes, etc. A *connector* is designed to connect the broken sections of a net. It can be implemented in different ways depending on the technology. Connectors for a bipolar VLSI are implemented by diffused tunnels, for instance. By *over-the-element route* we shall mean a connection which intersects the enclosing rectangle of an element (or a cell). The possibility of the construction such connections during circuit planarization is reflected in element models and can be ensured, for example, by the availability of areas within the rectangles where connections may be routed.

VLSI planarization is one of the basic stages (others will be discussed below) of the so called *topological* (in the mathematical sense) approach to VLSI design. This approach does not lie in the direction of the classical approach to automation of VLSI layout design. In the classical approach to computer-aided design the placement and routing problems are solved successively. The topological approach, in contrast, allows one to solve both problems at the same time. This is achieved by constructing a planar embedding of a circuit and obtaining the proper VLSI layout on the basis of it. Analysis of manual design techniques points to the fact that they are essentially based on the planarization process and its usage for layout construction. It is safe to say then that this method brings us back to the source.

In a broader sense, the topological approach assumes that the investigation and solution of the problems below rely on the same information.

1. Choice and construction of a mathematical model for layout design on the base of graph- and hypergraph-theoretic concepts.

2. Investigation of criteria and methods of planarizing graph and hypergraph models for circuits.
3. Development of metrization techniques, i.e. techniques for constructing an original geometric sketch of VLSI layout, on the basis of information about planarized circuit (embedding of a circuit).
4. Investigation and development of methods for optimizing an original metric sketch by different criteria (e.g. area) on the basis of layout deformation methods.
5. Investigation and development of computer geometry for testing and layout 'refinement'.

The classical approach to layout design automation has been widely presented in the literature, whereas its alternative is represented poorly (in spite of the tendency for an increase of studies on the subject). This may be attributed to an imperfect mathematical apparatus (graph planarization, in particular), sophisticated software, few such existing practical systems, which is virtually indicative of no feedback between the real practice and theory of system design and operation.

We expect the present monograph will clear up this difficulty to some extent. Although the work basically reflects our theoretical (mathematical) studies and experience in constructing and operating such layout design systems for the last 15 years, nevertheless it generalizes and presents the results of numerous studies in this field.

The theory and methods discussed here may be applied to printed-circuit boards, hybrid circuits, etc. It is no mere chance that the term 'VLSI' is used in the title. The matter is that the main difficulties encountered in solving the problems discussed herein are attributed to the necessity of overcoming the 'size' barrier. This also explains our invariable attention to estimating the efficiency of algorithms in question and the decomposition methods for large circuits discussed in Chapter 3. At the same time, the methods discussed here have been developed by the authors concurrently with the construction of an industrial computer-aided design system for the mass production of analog-digital VLSI.

The present work concentrates on 'essentially' hypergraph models of electric circuits and their planarization techniques. It is just this aspect of the topological approach to design that has not been adequately investigated.

Why hypergraph models? From the standpoint of the layout design process, the structure of a circuit can be represented most adequately in a hypergraph language. An electrical net connects the set of equipotential pins and has no *a priori* specified configuration. This set is to be treated essentially as a hyperedge in a circuit model. Moreover, physical planarization of a circuit performed by the designer is naturally modelled in the hypergraph language as a hypergraph planarization process, in the purely mathematical sense of the word. What is more, the proper notion of a planar hypergraph in its current formulation (as opposed to that of A.A. Zykov [92] where a hypergraph is said to be planar if it has a planar Konig representation) came about because of the similarity to circuit planarization mentioned. At the same time (hard-formalized) design techniques for layout area minimization (in particular, routing through an element, account of duplicate pins, etc.) is naturally reduced to a purely formal process of hypergraph planarization. It is as well to bear in mind that the hypergraph planarization problem is both interesting from the mathematical point of view and lies in the direction of classical investigation in planar graphs.

As to the other stages of the topological approach to VLSI layout design, we expect them to be covered in future works.

We shall now briefly run through the contents of the monograph.

The first three chapters deal with the fundamentals of the mathematical apparatus used in the subsequent discussion. These are first of all the fundamentals of graph and hypergraph theory, computational complexity of algorithms, and NP-completeness theory. Particular emphasis has been placed on the questions of graph and hypergraph planarity discussed in Chapters 2 and 3. There are many algorithms for graph planarity testing in the literature. Most efficient among them are the so called topological analysis algorithms of sequential embedding of a graph in a plane. Of this group Chapter 2 takes a closer look at two algorithms: the first has been proposed by Demoucron, Malgrange and Pertuiset, the second by Lempel, Even, and Cederbaum. Among the merits of the topological analysis algorithms is the possibility of taking them as a basis for constructing approximation algorithms to solve an, in practice, very important NP-complete problem of graph planarization, i.e. extraction of a maximum planar subgraph. Chapter 2 presents two such algorithms based on the above mentioned graph planarity testing algorithms.

Planarity can be extended to hypergraphs in different ways. The concept of a planar hypergraph discussed here is oriented to the solution of physical VLSI planarization problems. In contrast to the graph case, the problem remains NP-complete even in some special cases. Different classes of planar hypergraphs are discussed, the cases when the hypergraph planarity testing problem can be solved most efficiently are pointed out.

From the practical point of view, most important is the hypergraph planarization problem, i.e. finding the minimum number of hyperedge splits which create planarity of the resultant hypergraph, rather than just hypergraph planarity testing. Decomposition approaches can be used to best advantage both for hypergraph planarization and planarity testing, since they allow one to reduce the solution of an original problem to that of a collection of smaller problems for the so called 'non-decomposable' hypergraphs. Decomposition problems arising in hypergraph planarization are discussed in Chapter 3.

Chapter 4 deals with graph and hypergraph models of circuits which are adequate for the physical planarization problem. A distinction is made between a circuit model with prespecified models of elements and a circuit model for which there are no component models. The first type of model arises in VLSI layout design on the basis of earlier designed layouts of elements or cells. The second type of model arises in top-down layout design when the layout design conditions for elements or cells and hence their models are determined dynamically in the course of VLSI layout constructing.

Chapter 5 is concerned with the extraction and embedding of a maximum planar part of a VLSI. The problem is reduced to that of hypergraph planarization. The decomposition planarization methods of Chapter 3 allow one to reduce the solution of an arbitrary hypergraph planarization problem to that of a similar problem for 'non-decomposable' hypergraphs. Chapter 5 proposes two approaches to the approximate solution of such problems.

One is the iteration method which determines and embeds some part of a hypergraph (the so called segment) at each step. In this regard the stated efficient criterion of segment embeddability takes on great significance. The application of this method to graph planarity testing leads us to the well-known Demoucron, Malgrange and Pertuiset algorithm.

The second approach consists in constructing on the basis of an original hypergraph G a hypergraph G' with the planar Konig representation that has been obtained by splitting vertices of graph $K(G)$ which correspond to edges of G. Generally, the construction of the graph $K(G')$ by the minimum number of splits is an NP-complete problem. An approximation algorithm based on that of Lempel, Even and Cederbaum has been proposed as its solution.

Chapter 6 discusses the problems of constructing nonplanar connections. To construct an unembedded connection of a circuit one can use both over-the-element routes and connectors. The mathematical model for this problem is a special nonplanar connection graph introduced in Chapter 6.

Constructing a nonplanar connection is reduced to the problem of constructing in the nonplanar connection graph a minimal tree whose special case is a well-known NP-complete problem of constructing the Steiner tree in a weighted graph. Approximation algorithms for solving the problem are presented.

The order of constructing nonplanar connections has a pronounced effect on the number of required connectors, their length, etc. Chapter 6 deals with selecting the order of nonplanar connections which minimizes the total length of connectors. Approximation algorithms for the determination of the order have been proposed.

Chapter 7 is concerned with the structure of a VLSI planarization system based on the circuit planarization methods that have been discussed earlier. With due regard of some topological and metric constraints, the system:

- tests a circuit for planarity;
- extracts and embeds in the plane the maximum planar part of a circuit;
- constructs nonplanar connections by placing connectors and over-the-element routes;
- minimizes the number of circuit elements over which connections are routed;
- minimizes the total length of connectors.

Chapter 1

Discrete Mathematics Fundamentals

1.1. Graphs And Hypergraphs

A *graph* (undirected) can be defined as a pair (V, E) where V is a non-empty finite set and E is a family of distinct unordered pairs of distinct elements of V. Elements of V are called *vertices* and elements of E are called *edges*. A graph is normally represented schematically as points and lines joining some of these points. The points are associated with vertices of a graph, whereas, lines joining pairs of points are associated with its edges. Figure 1.1 shows a graph with 5 vertices and 6 edges, with $\{1,2,3,4,5\}$ being its vertex set and $\{\{1,2\},\{2,3\},\{3,4\},\{3,5\},\{1,5\}\}$ being a family of its edges. The number of vertices in a graph is called the *order* of the graph.

The above definition of a graph implies that all elements of E are distinct. Assuming the existence of equal elements in E we come to a concept of a *multigraph*. Distinct elements of E are then called *multiple edges*. Admitting there is a pair of the form $[v,v]$ in E we come to a concept of a *pseudograph*. The pair is called a *loop*. Figure 1.2 shows a multigraph and a pseudograph. The vertices u and v are called the *endpoints* of the edge $e = [u,v]$ and e is said to join the vertices u and v. Two edges are said to be *adjacent* if they have a common endpoint. A vertex u and edge e are *incident* if u is the endpoint of e; otherwise they are *non-incident*.

Directed graphs (digraphs) are also under study nowadays. A *digraph* is a pair (V, A) where V is a non-empty finite set and A is a set of ordered pairs of distinct elements of V. Elements of V are called its *vertices* and elements of A are called *directed edges* or *arcs*. If $a = (u,v)$ is an arc then u is called its *initial endpoint* and v its *terminal endpoint*. In the figure the arcs are represented by arrows showing the direction from initial endpoint to terminal endpoint (Figure 1.3). As with the undirected case, directed multigraphs and

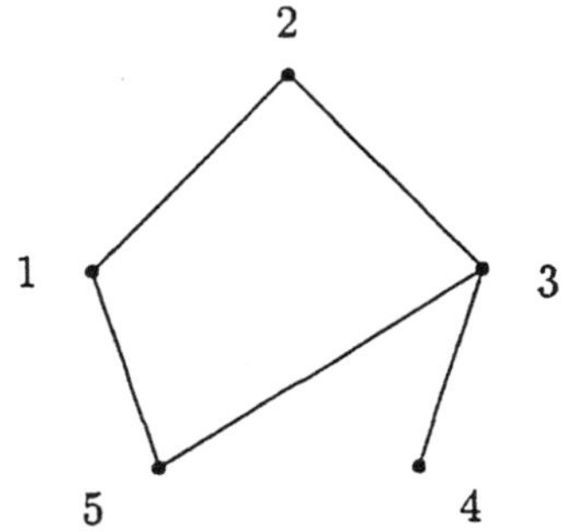

Fig. 1.1. A graph with 5 vertices and 6 edges.

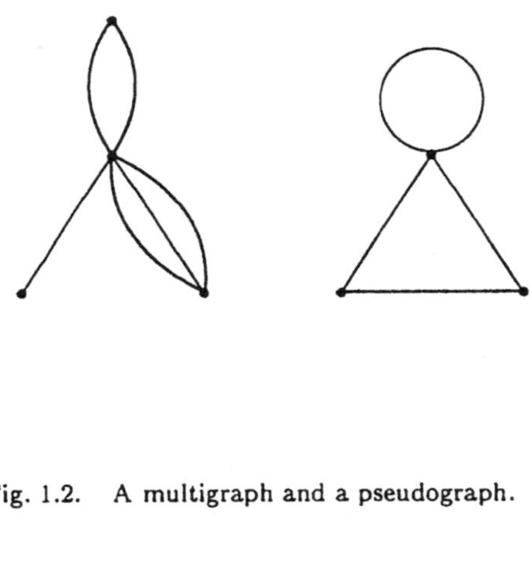

Fig. 1.2. A multigraph and a pseudograph.

pseudographs can be defined (Figure 1.4).

If a number $w(e)$ (called the *weight*) is associated with each edge e of a graph (or each arc e of a digraph) we will treat it as a *weighted graph* (or *weighted digraph*).

A graph $G_1 = (V_1, E_1)$ is called the *subgraph* of $G = (V, E)$ if $V_1 \subseteq V$ and $E_1 \subseteq E$. In this case G is called a *supergraph* of G_1. G_1 is called a *spanning subgraph* if $V_1 = V$. If E_1 coincides with the set of all edges of E which are only incident with the vertices of V_1, G_1 is said to be the *subgraph induced by* V_1, and is denoted by $G(V_1)$. There are also *subgraphs induced by the edge sets* which are under study. For $E_1 \subseteq E$, the edge set of a induced subgraph $G(E_1)$

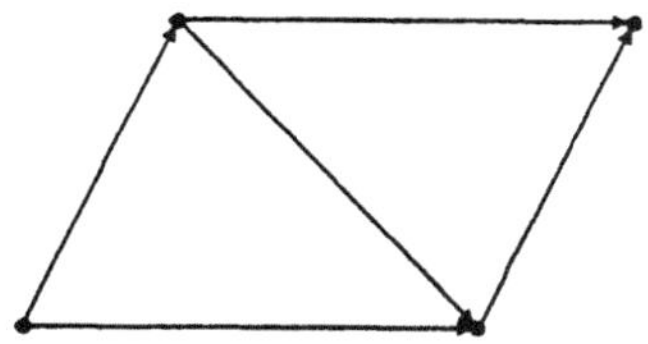

Fig. 1.3. A digraph.

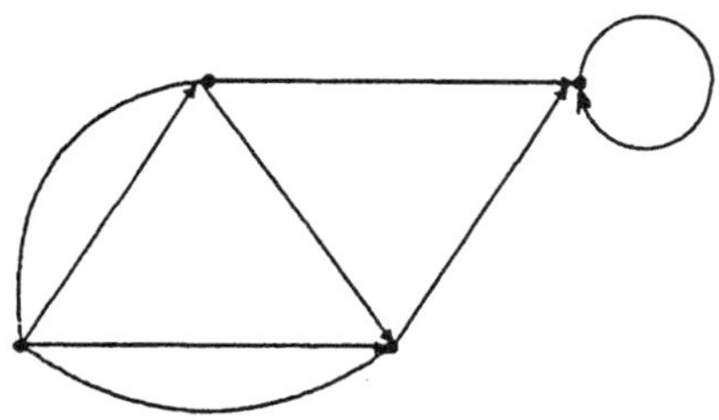

Fig. 1.4. A directed pseudograph.

coincides with E_1, and the vertex set coincides with the set of endpoints of the edges from E_1. Figure 1.5 shows a graph G and its three subgraphs G_1, G_2 and G_3, among which G_2 is the spanning subgraph and G_3 is induced by the vertex set. In the subsequent discussion of induced subgraphs we will mean a subgraph induced by a vertex subset.

Since conventional graph-theory models and methods usually serve as apparatus for describing mainly binary relations, they fail to describe adequately many problems of practical importance. A major feature of such problems is the necessity of describing various types of n-ary relations. A new branch of graph theory dealing with the hypergraphs opens up fresh opportunities in that respect.

A hypergraph is a generalized notion of a graph whose edges may be any non-empty vertex subsets rather than a 2-element subset. All basic definitions of graph theory can easily extend to hypergraphs, therefore we will give them immediately for hypergraphs. For a reader who wishes to take a close look at

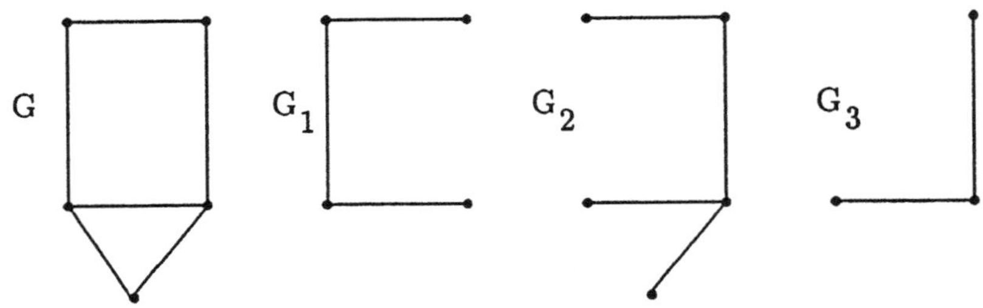

Fig. 1.5. A graph and its subgraphs.

graph and hypergraph theory we recommend the following literature [7, 32].

A *hypergraph* is a pair $(V, \mathcal{H})$, where V is a non-empty finite set and $\mathcal{H}$ is a family of non-empty subsets of V. The elements of V are called its *vertices* and the elements of $\mathcal{H}$ are called its *(hyper)edges*. Equal subsets in $\mathcal{H}$ are called *multiple(hyper)edges*. The number of vertices in G is called its *order*, which is denoted by $|G|$.

A vertex $v \in V$ and edge $H \in \mathcal{H}$ are said to be *incident* if $v \in H$. We denote by $\mathcal{H}(v)$ the set of all edges which are incident with the vertex $v \in V$. The number $|\mathcal{H}(v)|$ is called the *degree* of vertex v, and the number $|H|$ is called the *degree* of edge H. The maximal degree of an edge in G is called the *rank* of the hypergraph G and is denoted by *rankG*. A vertex of a hypergraph which is not incident with any edge is said to be *isolated*. A vertex of degree 1 is called *pendant*. Two vertices are *adjacent* if there is an edge incident with both vertices. We shall call two edges H_1 and H_2 adjacent if $H_1 \cap H_2 \neq \emptyset$.

In the figure edge H is drawn as a solid line which encircles all the vertices of H if $|H| > 2$ or $|H| = 1$. If $|H| = 2$, such an edge will be shown the same as in the graph case as a line joining its two vertices. Figure 1.6 shows the hypergraphs with the vertex set $V = \{v_1, v_2, ..., v_9\}$ and edge set $\mathcal{H} = \{H_1 = \{v_1, v_2, v_3\},\ H_2 = \{v_4, v_5, v_8\},\ H_3 = \{v_3, v_4\},\ H_4 = \{v_2, v_5, v_7, v_9\},\ H_5 = \{v_8, v_9\},\ H_6 = \{v_2, v_7\}\}$.

A hypergraph $G_1 = (V_1, \mathcal{H}_1)$ is said to be a *partial hypergraph* (or *part*) of the hypergraph $G = (V, \mathcal{H})$ if $V_1 \subseteq V$ and $\mathcal{H}_1 \subseteq \mathcal{H}$. The hypergraph $(V_1, \mathcal{H}_1)$, where $\mathcal{H}_1 = \{H \in \mathcal{H} : H \subseteq V_1\}$, is called the *partial hypergraph* of $G = (V, \mathcal{H})$ *induced by the vertex set* V_1. The hypergraph $(V_1, \mathcal{H}_1)$ where $\mathcal{H}_1 = \{H' = H \cap V_1 \neq \emptyset : H \in \mathcal{H}\}$ is called the *subhypergraph* of $G = (V, \mathcal{H})$ *induced by the vertex set* V_1.

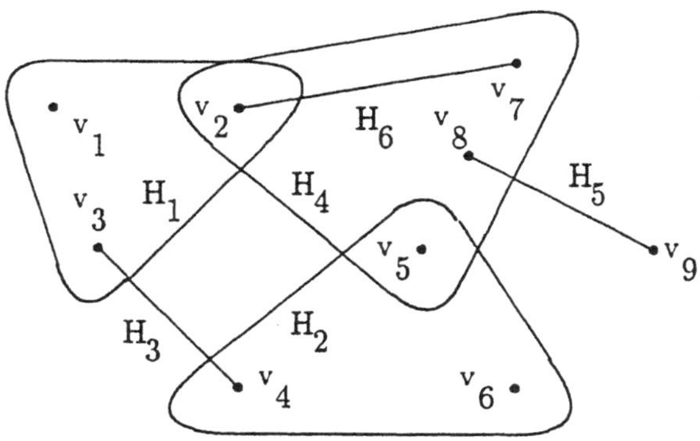

Fig. 1.6. A hypergraph.

If the degrees of all edges of a hypergraph are k the hypergraph is said to be k-uniform. Clearly, a graph is a 2-uniform hypergraph. Then a graph is a special case of hypergraph, and therefore all definitions relating to hypergraphs hold true for graphs.

A hypergraph is called *k-regular* if the degrees of all its vertices are k. 3-regular hypergraphs are called *cubic*.

Any hypergraph $G = (V, \mathcal{H})$ without isolated vertices can be associated with a hypergraph $G^* = (V^*, \mathcal{H}^*)$ whose vertex set V^* coincides with $\mathcal{H}$ and $\mathcal{H}^*$ is the family of all sets $\mathcal{H}(v), v \in V$. The hypergraph G^* is called the *dual hypergraph* of G. Figure 1.7 shows a hypergraph which is the dual hypergraph of that of Figure 1.6.

In the subsequent discussion we assume that the degrees of all edges in a hypergraph are greater than 1 and there are no multiple edges in the hypergraph.

For an arbitrary hypergraph $G = (V, \mathcal{H})$ we define a graph $K(G)$. The vertex set of this graph coincides with the set $V \cup \mathcal{H}$. Two vertices $v \in V$ and $H \in \mathcal{H}$ are adjacent if $v \in H$. The graph $K(G)$ is called a *Konig representation* of hypergraph G. Figure 1.8 shows the Konig representation of hypergraph G of Figure 1.6.

In a hypergraph $G = (V, \mathcal{H})$ a *chain of length* $l > 0$ is defined to be a sequence

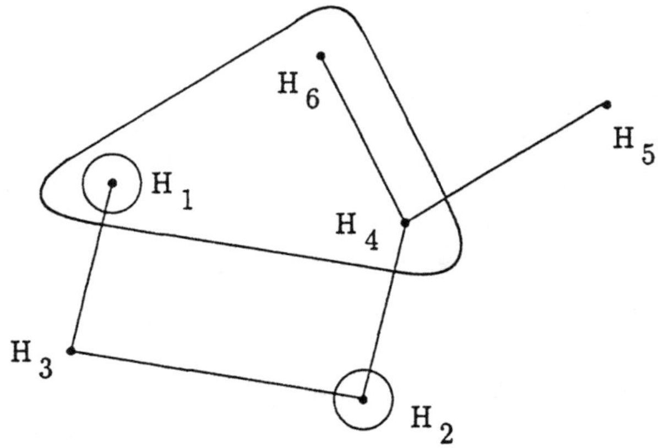

Fig. 1.7. The dual hypergraph of that of Figure 1.6

$$v_1, H_1, v_2, H_2, ..., H_l, v_{l+1}$$

of vertices and edges of hypergraph such that

1. all vertices $v_1, ..., v_{l+1}$ are distinct, except possibly the initial and terminal ones;
2. $H_1, ..., H_l$ are distinct edges of $\mathcal{H}$;
3. $v_i, v_{i+1} \in H_i$ for any $i = 1, ..., l$.

The chain connecting the vertices v_i and v_{l+1} is called a (v_i, v_{l+1})-*chain.*

If $l > 1$ and $v_{l+1} = v_1$ the chain is called a *cycle* . The cycle which contains all vertices of a hypergraph G is called a *Hamilton cycle.*

The vertices u and v of hypergraph G are said to be *connected* if $u = v$ or a (u, v)-chain exists in G. Clearly, the connectivity relation is an equivalence relation on the vertex set of a hypergraph and divides the vertex set into classes of pairwise connected vertices which are called *connected domains.* Subhypergraphs induced by the connected domains are called *connected components.* If there is one connected component in a hypergraph the hypergraph is called *connected.*

In a digraph $G = (V, A)$ a *path of length* $l > 0$ is defined to be a sequence

$$v_1, a_1, v_2, a_2, ..., a_l, v_{l+1}$$

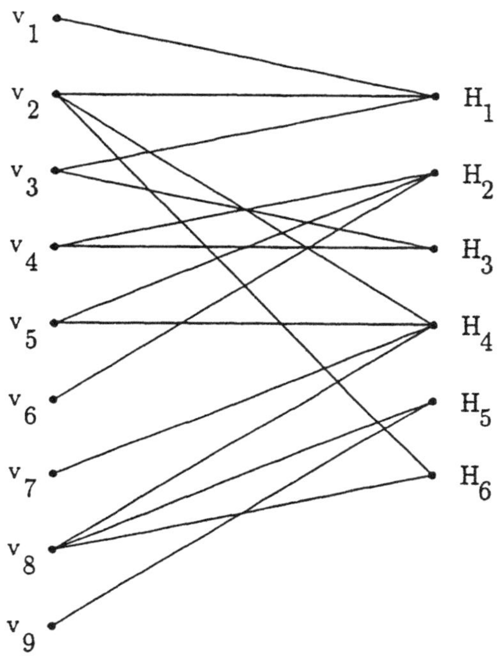

Fig. 1.8. Konig representation of the hypergraph of Figure 1.6

of vertices and arcs of digraph such that

1. all vertices $v_1, ..., v_{l+1}$ are distinct except possibly the initial and terminal ones;
2. $a_1, ..., a_l$ are distinct arcs of A ;
3. $a_i = (v_i, v_{i+1})$ for any i=*1,...,l*.

The path connecting the vertices v_i and v_{l+1} is called a (v_i, v_{l+1})-*path*. If $l > 1$ and $v_{l+1} = v_1$ the path is called a *cycle*. A vertex v is called *reached* from a vertex u if there exists a (u, v)-path.

A *line graph* $L(G)$ of a hypergraph G is the graph whose vertex set coincides with the edge set of G. Two vertices are adjacent if their corresponding edges are adjacent.

Consider operations on hypergraphs.

An operation of transition from a hypergraph G to a hypergraph G_v which is a subhypergraph of G induced by the vertex set $V \backslash \{v\}$ is called *removal of a vertex* $v \in V$ *from hypergraph* $G = (V, \mathcal{H})$. An operation of transition from the original hypergraph $G = (V, \mathcal{H})$ to the hypergraph $(V, \mathcal{H}\backslash\{H\})$ is called *removal of the edge* $H \in \mathcal{H}$.

An operation of transition from the hypergraph $G = (V, \mathcal{H})$ to the hypergraph $G' = (V', \mathcal{H}')$ where $V' = V \cup \{v\}, \mathcal{H}' = (\mathcal{H}\backslash\{H\}) \cup \{H \cup \{v\}\}$ is called *insertion of the vertex* v *into edge* H

An operation resulting in the removal of all vertices which are incident with H and the addition of a new vertex which is inserted into each edge that has been earlier adjacent to H is called *contraction of the edge* H *in the hypergraph* G. Put another way, the contraction of the edge H is a transition from the hypergraph $G = (V, \mathcal{H})$ to the hypergraph $G' = (V', \mathcal{H}')$ where

$$V' = (V\backslash H) \cup \{u_H\}$$

$$\mathcal{H}' = \mathcal{H}_1 \cup \mathcal{H}_2$$

$$\mathcal{H}_1 = \{H' \in \mathcal{H} : H' \cap H = \emptyset\}$$

$$\mathcal{H}_2 = \{(H'\backslash H) \cup \{u_H\} : H' \in \mathcal{H}, H' \cap H \neq \emptyset\}.$$

The *union* of two hypergraphs $G_1 = (V_1, \mathcal{H}_1)$ and $G_2 = (V_2, \mathcal{H}_2)$ is a hypergraph

$$G_1 \cup G_2 = (V_1 \cup V_2, \mathcal{H}_1 \cup \mathcal{H}_2).$$

The *difference* $G_1\backslash G_2$ of two hypergraphs $G_1 = (V_1, \mathcal{H}_1)$ and $G_2 = (V_2, \mathcal{H}_2)$ is a subhypergraph of the hypergraph $(V_1, \mathcal{H}_1\backslash\mathcal{H}_2)$ generated by the vertex set $V_1\backslash V_2$.

Consider the examples of some special graph classes.

A graph G is *complete* if any pair of its vertices is adjacent. The complete graph of order n is denoted by K_n. The number of edges in it is $n(n-1)/2$. Figure 1.9 shows the graphs K_4 and K_5.

A graph is called *empty* if it has no edges. The empty graph of order n is denoted by O_n.

A graph is *bipartite* if its vertex set can be divided into two parts such that the vertices belonging to the same part are not adjacent. Thus in a bipartite graph the endpoints of any edge belong to different parts. Moreover, if any two vertices belonging to different parts are adjacent the graph is called *complete bipartite.* The complete bipartite graph whose parts contain n and m vertices is denoted by $K_{n,m}$. With $n = 1$ a *star* $K_{1,m}$ is obtained.

Figure 1.10 shows the star $K_{1,3}$ and the complete bipartite graph $K_{3,3}$.

Denote the cycle of length n (Figure 1.11) by C_n and the chain of length $n-1$ (Figure 1.12) by P_n. A graph containing no induced subgraph $C_n, n > 3$

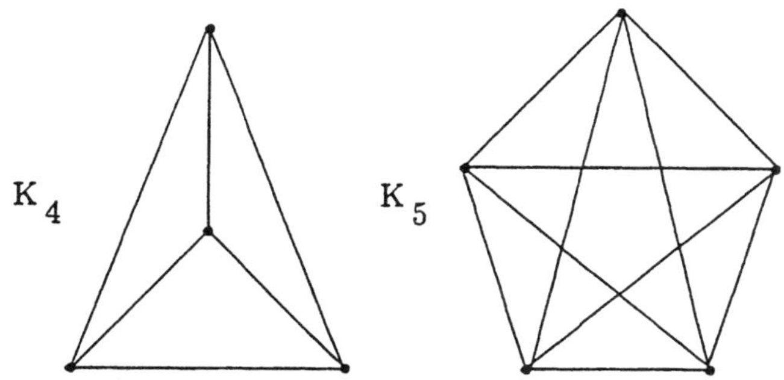

Fig. 1.9. The graphs K_4 and K_5

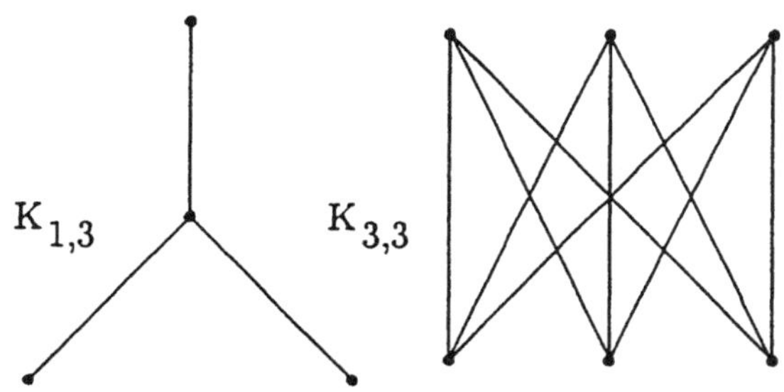

Fig. 1.10. The star $K_{1,3}$ and the complete bipartite graph $K_{3,3}$

is called *chordal.* The union of graph C_n and star $K_{1,n}$ is called a *wheel* which is denoted by W_n (Figure 1.13).

A *tree* is a connected graph which has no cycles. The following conditions are equivalent for a graph G with n vertices and m edges:

1. G is a tree;
2. G is a connected graph and $m = n - 1$;
3. G is a graph without cycles and $m = n - 1$;
4. any two distinct vertices of G are connected by a unique chain;
5. G is a graph without cycles with the property that if any pair of non-adjacent vertices of the graph is connected with an edge the resulting graph will contain a single cycle.

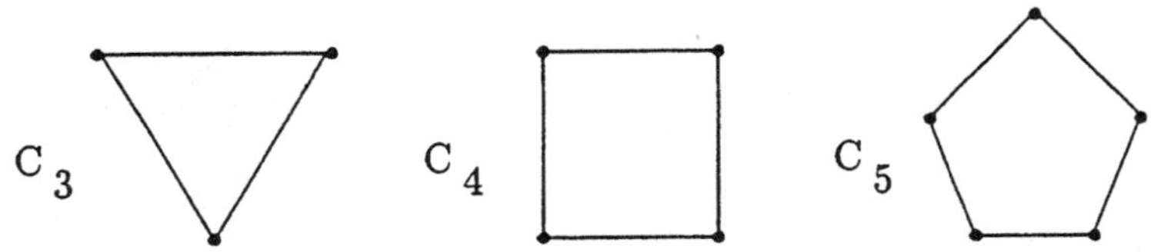

Fig. 1.11. Cycles of length 3, 4 and 5

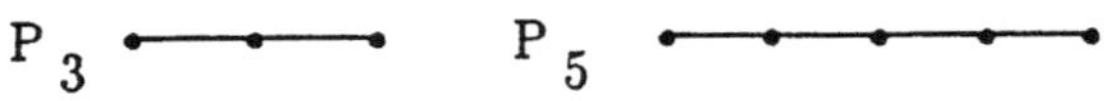

Fig. 1.12. Chains of length 3 and 5

A *directed tree* (or a *root tree*) is a digraph without cycles which satisfies the following conditions:

1. there is a unique vertex (called the *root*) which is not a terminal endpoint of any arc;
2. any vertex, except the root, is a terminal endpoint of only one arc;
3. there is a path from the root to any other vertex.

If $G = (V, A)$ is a root tree and $(u, v) \in A$ then v is called a *child* of u and u is called a *parent* of v. A vertex which is not an initial endpoint of any arc is called a *leaf*.

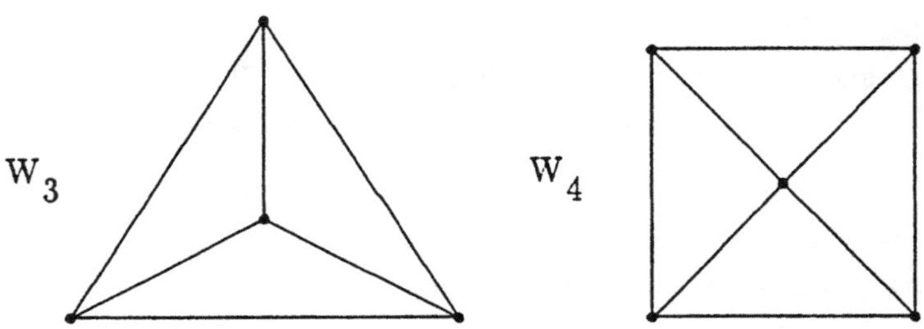

Fig. 1.13. Wheels W_3 and W_4

Now we shall define some invariants of hypergraphs.

The vertex set U of a hypergraph $G = (V, \mathcal{H})$ is *independent* if U contains no edges from $\mathcal{H}$. The maximum cardinality of an independent set of G is called the *independence number* denoted by $\alpha_0(G)$.

A *k-colouring* of a hypergraph $G = (V, \mathcal{H})$ is an arbitrary function $f : V \rightarrow \{1, ..., k\}$. A k-colouring can be treated as colouring each of the hypergraph vertices into one of k colours. A *proper k-colouring* of a hypergraph is a colouring in which no adjacent vertices belonging to an edge are the same colour. The minimum number of colours required for proper k-colouring of G is called the *chromatic number* and is denoted by $\chi(G)$. If $\chi(G) = k$ the hypergraph G is called *k-chromatic.*

The *vertex connectivity* of a connected hypergraph G is the minimal number of vertices which, when removed, makes the hypergraph G either unconnected or a single vertex. The vertex connectivity of a hypergraph G is denoted by $\kappa(G)$. If $\kappa(G) \geq k, G$ is *k-connected.*

A vertex v is called an *articulation vertex* of the hypergraph if the number of connected components increases with its removal.

The maximal connected subhypergraph without any articulation vertex is called a *block.* Clearly, the 2-connected hypergraph is a block in itself.

An *articulation set of vertices* (*edges*) of a connected hypergraph is the set of its vertices (edges) which, when removed, causes the hypergraph to become unconnected.

The minimal articulation set of edges is called a *cut* (or a *cutset*). In a hypergraph the edge which is a cut is called a *bridge.* The *edge connectivity* of a connected hypergraph is the minimum number of edges in the cut. The edge connectivity of a hypergraph G is denoted by $\lambda(G)$. If $\lambda(G) = k, G$ is said to be *k-edge-connected.*

We can show [13] that

$$\lambda(G) \leq \delta(G)$$

$$\kappa(G) \leq (rank G - 1)\lambda(G)$$

where $\delta(G)$ is the minimum degree of a vertex of G.

1.2. Algorithms and Their Complexity

There are different criteria for estimating algorithms. A major one is the time required for running the appropriate programme on a computer. Clearly, this factor essentially depends on the type of computer in use. To make our conclusions more general, consider an abstract computer.

Similarly to [91], such a computer can perform arithmetic operations, comparisons, moves, conditional and unconditional branches. These operations are considered elementary. An elementary operation running time is 1. Thus the algorithm running time is the number of elementary operations performed by the algorithm.

The storage of our abstract computer consists of an unlimited number of random-access locations. Consider the representation of numbers in the computer memory. When analysing the majority of algorithms presented in this book it would be convenient to abstract the value of many numeric parameters (for example, graph edge weights). Therefore we shall consider that any number, irrespective of its length, can be placed in one location, and that any location can contain only one number. In item 1.4 we shall slightly change the adopted computer model, in particular, a different way of number representation.

Consider now the notion of an input length. For combinatorial problems, any combinatorial object, i.e. a graph, a hypergraph, a family of finite sets, etc., may be used as ***input data*** (or ***input***). For the data to be entered into a computer it must be represented as a sequence of some characters in a certain fashion. These representations can be performed in many evident and straightforward ways; below are some ways of coding graphs and hypergraphs. However, all these representations, almost without exception, appear to be equivalent for our present objectives. Thus the algorithm input is represented as a sequence of characters. The ***input length*** is the number of characters in the sequence.

Consider, then, the representation of a hypergraph $G = (V, \mathcal{H})$ in the computer storage. Let $V = \{v_1, ..., v_n\}$ and $\mathcal{H} = \{H_1, ..., H_m\}$. The hypergraph can be represented in several ways.

One of them is to represent a hypergraph by an incidence matrix. The *incidence matrix* is the $n \times m$ matrix $I = \parallel I_{ij} \parallel$ where

$$I_{ij} = \begin{cases} 1 & \text{if } v_i \in H_j, \\ 0 & \text{if } v_i \notin H_j. \end{cases}$$

Clearly, $O(nm)$ storage locations are required to allocate the hypergraph.

If G is a graph there may normally be three ways of its representation.

First, the graph can be specified by the adjacency matrix. The ***adjacency matrix*** is the $n \times m$ matrix $A = \parallel A_{ij} \parallel$ where

$$A_{ij} = \begin{cases} 1 & \text{if vertices } v_i \text{ and } v_j \text{ are adjacent,} \\ 0 & \text{otherwise.} \end{cases}$$

For a weighted graph the adjacency matrix is substituted with a weight $n \times m$ matrix $W = \parallel W_{ij} \parallel$ where $W_{ij} = W(v_i, v_j)$ if v_i and v_j are adjacent and $W_{ij} = 0$ or $W_{ij} = \infty$ (depending on the problem); otherwise, with $W(v_i, v_j)$ being the weight of edge $[v_i, v_j]$. $O(n^2)$ storage locations are required for such a representation of the graph.

The second way is to specify the graph as a list of edges. Since any edge of the graph can be stored in two locations (one location per endpoint). $O(2m)$ locations will be sufficient to allocate the entire list.

The third way is to specify the graph as an adjacency list where each vertex is associated with the list of adjacent vertices.

The graph representation techniques described above can be used to represent a hypergraph through its Konig representation. The choice of representation technique depends on the problem to be solved.

When analysing the algorithm we shall primarily be interested in the relationship between the running time and the input length. The *time complexity function* (or *time complexity*) of an algorithm is the function f which associates each natural number n with the worst-case running time $f(n)$ of the algorithm for inputs of length n. In other words, $f(n)$ is the maximum running time for the algorithm at all inputs of length n. To estimate the time complexity for the algorithm, the O-notation is often used. We will write $f(n) = O(g(n))$ if there exists a constant c such that $f(n) < cg(n)$ for all $n > 0$. Algorithms with time complexity $O(n)$ are called *linear* and those with time complexity $O(p(n))$, where $p(n)$ is a polynomial of n, are called *polynomial.*

1.3. Algorithms In Graphs

1.3.1. *Depth-first search*

Depth-first search is a way of running through all vertices and edges of a graph. It underlies the development of many fast algorithms in graphs.

Let $G = (V, E)$ be an undirected connected graph. During a depth-first search the vertices of graphs are numbered and the edges are labelled.

Select an arbitrary vertex v_0 in G. Assume that all vertices of G are unnumbered and the edges are unlabelled. Assign the number 1 to vertex v_0. Select an arbitrary edge $[v_0, u]$, label it 'direct' and assign the number 2 to vertex u and go to vertex u. Assume that some steps of this process result in arriving at vertex w and let k be the last number assigned to vertices. Then proceed as follows.

Let there exist an edge $[w, v]$. If vertex v is already numbered, label the edge as 'reverse' and continue searching for an unlabelled edge which is incident with vertex w. If vertex v is not yet numbered assign the number $k+1$ to it, label edge $[w, v]$ as 'direct' and go to vertex v. The vertex v will be considered as numbered from w. If all edges that are incident with w have already been labelled, go to the vertex wherefrom vertex w has been numbered. The process is completed when all edges are labelled and vertex v_0 is reached again. The depth-first search algorithm in an undirected graph looks as follows.

Depth-First Search Algorithm

Step 1. Given a graph $G = (V, E)$. Let $T = \emptyset$, $B = \emptyset$, $k = 1$. Put $parent(v) = 0$ and $mark(v) = 0$ for any vertex v.

Step 2. Select an arbitrary vertex v_0 of G for which $mark(v_0) = 0$. Let $mark(v_0) = k$, $v = v_0$.

Step 3. If all edges incident with v are labelled go to step 4. Otherwise, select an unlabelled edge $[v, w]$ and label it. If $mark(w) = 0$ then $k = k+1$, $T = T \cup \{[v, w]\}$, $mark(w) = k$, $parent(w) = v$, $v = w$. But if $mark(w) \neq 0$ assume that $B = B \cup \{[v, w]\}$. Go to step 3.

Step 4. If $parent(v) \neq 0$, assume $v = parent(v)$ and go to step 3. Otherwise, go to step 5.

Step 5. If $mark(u) \neq 0$ for all vertices u the algorithm is completed. Otherwise, assume that $k = k+1$ and go to step 2.

The time complexity of this algorithm does not exceed $O(m)$, where m is the number of edges in a graph. The list T of 'direct' edges that has been constructed during the depth-first search generates a spanning tree of G.

The depth-first search can be used for finding the connected components of a graph.

1.3.2. *Breadth-first search*

Another way of running through all vertices and edges of a graph is breadth-first search. It consists in the following.

Select an arbitrary vertex v_0 and assign it number 0. Any vertex adjacent to v_0 is assigned the number 1. Then each unnumbered vertex adjacent to vertex with number 1 is assigned the number 2. Then consider the vertices adjacent to that with number 2, etc, as long as is possible. It would be suitable to build a list S for this algorithm. Originally, S contains only one vertex v_0. The vertex is assigned the number $l(v_0) = 0$. The iteration is as follows. Select the first vertex v from S. For each unlabelled vertex u adjacent to v assume that $l(u) = l(v)+1$. Vertex u is placed at the end of S and vertex v is removed from the list.

Breadth-First Search Algorithm

Step 1. $S(1) = v_0$, $i = 1$, $j = 1$, $l(v_0) = 0$.

Step 2. $v = S(i)$. Let N_v be the list of vertices adjacent to v and p is the length of the list.

Step 3. $k = 1$.

Step 4. If vertex $u = N_v(k)$ is already numbered, go to step 5. Otherwise, assume that $l(u) = l(v) + 1$, $j = j + 1$, $S(j) = u$ and go to step 5.

Step 5. If $k = p$, go to step 6. Otherwise, let $k = k + 1$ and go to step 4.

Step 6. $i = i + 1$.

Step 7. If $i > j$ the algorithm is completed. Otherwise, go to step 2.

The time complexity of this algorithm is $O(m)$, where m is the number of edges in the graph. For a connected graph the breadth-first search labels all vertices. Thus the algorithm can be used for analysing the connectivity of a graph.

Besides this, the number $l(v)$ assigned to vertex v during breadth-first search indicates the length of the shortest chain connecting v_0 and v. To identify the chain, a vertex from which the given vertex v has been labelled must be kept for every vertex v.

1.3.3. *Minimum spanning tree*

The minimum spanning tree problem consists in finding a minimum weight spanning tree in the weighted graph. We describe the application of Kruskal's algorithm for solving the problem.

Let $G = (V, E)$ be an undirected weighted graph. The algorithm consists in the sequential addition of minimum weight edges to the graph $T = (V, \emptyset)$ such that T contains no cycles. The execution of the algorithm results in a minimum weight spanning tree T. When describing the algorithm we shall denote by $\mathcal{K}$ the set of connected domains of a graph T.

Kruskal's Algorithm

Step 1. $T = (V, \emptyset)$, $\mathcal{K} = \{\{v\} : v \in V\}$. Sort edges from E in the increasing order of their weights: $E = \{e_1, ...e_m\}$; $i = 1$.

Step 2. If the endpoints of an edge e_i belong to different subsets $V^{'}$ and $V^{"}$ from $\mathcal{K}$, add edge e_i to graph T, substitute V' and $V^{"}$ for $V^{'} \cup V^{"}$ and setting $i = i + 1$ go to step 3. Otherwise, setting $i = i + 1$, go to step 2.

Step 3. If $|\mathcal{K}| > 1$ go to step 2. Otherwise, the algorithm running completes.

The time complexity of this algorithm is $O(m \log m)$.

1.3.4. *The shortest chain*

Let $G = (V, E)$ be an undirected, connected, weighted graph. We shall find a minimum weight chain connecting the two given vertices s and t. Consider the case when the weights of all edges are non-negative, i.e. $w(e) \geq 0$ for all $e \in E$.

Consider the application of Dijkstra's algorithm for solving this problem. On each iteration of the algorithm, any vertex v of a graph has been assigned label $l(v)$ which is either permanent or temporary. The permanent label $l(v)$ shows the weight of the shortest (s, v)-chain. The temporary label $l(v)$, however, shows the weight of the shortest (s, v)-chain which passes through the vertices with permanent labels. Moreover, every vertex v except s has a label $p(v)$. On each iteration, $p(v)$ is the vertex preceding v in the (s, v)-chain the weigth of which is minimum among all (s, v)-chains with vertices having only permanent labels l. By using labels $p(v)$, we can readily show the sequence of vertices which form the shortest (s, v)-chain.

Before the first iteration starts, the vertex s has a permanent label $l(s) = 0$ and the labels of all the remaining vertices are temporary and equal to ∞. The general iteration consists in the following. Let v be a vertex which has been assigned a permanent label in the preceding iterations. Observe all vertices u with temporary labels which are adjacent to v to reduce the labels.

If $l(u) > l(v) + w(v,u)$ assume $l(u) = l(v) + w(v,u)$, $p(u) = v$. Otherwise, labels l and p remain unchanged for vertex u. The algorithm stops running when label $l(t)$ becomes permanent.

Dijkstra's Algorithm

Step 1. Let $v = s$, $l(s) = 0$ and assume the label is permanent. Let $l(v) = \infty$ for all $v \in V \backslash \{s\}$ and assume the labels be temporary.

Step 2. For every v with the temporary label which is adjacent to v, perform: if $l(u) > l(v) + w(v.u)$ then $l(u) = l(v) + l(v,u)$ and $p(u){=}v$.

Step 3. Among all vertices with temporary labels find a vertex v^* with the minimal value $l(v^*)$. Label $l(v^*)$ is considered to be permanent.

Step 4. $v = v^*$. If $v \neq t$, go to step 2. Otherwise, $(s, ..., p(p(t)), p(t), t)$ is the shortest (s,t)-chain and $l(t)$ is the weight of the chain.

The time complexity of this algorithm is $O(n^2)$, where n is the number of vertices of a graph.

1.4. Intractable Problems

An algorithm is commonly considered efficient if its time complexity is polynomial. However a polynomial algorithm for problem solving can be constructed far from every problem of practical importance. This item discusses the problems of such a class. We adopt the approach of [91].

We shall introduce some modifications in the computer model of item 1.2. Assume that every location can contain only 0 or 1 and that the integers are represented in binary notation. Thus to represent an integer $K \geq 0$, we require $]\log|K|[$ storage locations ($]x[$ is the smallest integer which is not less than x). A rational number is treated as an irreducible fraction and represented as an ordered pair of integers: the numerator and the denominator of the fraction. An elementary operation time is the sum of record lengths for its operands in binary.

In what follows we consider each problem in the so-called *recognition form* when the problem solution consists in getting an answer 'yes' or 'no'. For example, is the given hypergraph planar? In practice, however, the initial statement of a problem is normally optimization, i.e. a solution x whose weight $w(x)$(or cost) is minimum must be selected among the set X of admissible solution. Every optimization problem of this kind can be associated with its recognition form, which is as follows. Given a set X, a weight function w and a number K, we must determine if there is an element $x \in X$ such that $w(x) \leq K$. Clearly, if there exists a polynomial algorithm for solving the initial optimization problem it also exists for solving the associated recognition (or decision) problem. Moreover, with slight constraints on the weight function w the reverse is also true. We shall not elaborate on the details here but only note that the optimization problem is 'not more simple' than the associated recognition problem.

We denote by P a class of recognition problems each of which can be solved by the algorithm in polynomial time. Specifically, the minimum spanning tree and shortest chain problems discussed above (their recognition forms naturally) belong to class P. Let us define another class NP of recognition problems with the property that if the instance of the problem has a "yes" answer there exists a polynomial algorithm which verifies (proves) the fact. Clearly $P \subseteq NP$. The question whether the sets P and NP coincide has remained open up to now.

Next consider the concept of problem transformation. Generally, the *transformation of a problem A to a problem B* means that the solution of problem A can be obtained from the solution of problem B. This, however, does not account for the computational complexity of obtaining solution of problem A from that of problem B. If a polynomial algorithm can be used for this purpose we talk about *polynomial transformation of A to B*. The term 'polynomial' will be omitted hereafter. In other words, problem A is polynomially transformed to problem B ($A \propto B$) if there is a polynomial algorithm which creates instance I' of B for any instance I of A and answers 'yes' if and only if 'yes' has input I'. Clearly if $B \in P$ and $A \propto B$, then $A \in P$; if $A \propto B$, $B \propto C$, then $A \propto C$.

Problem A is said to be *NP-complete* if $A \in NP$ and any problem from NP is transformed to A. In other words, NP-complete problems are the 'hardest' problems of NP. A problem is called *NP-hard* if an NP-complete problem can be transformed to it. Below are the examples of some NP-complete problems.

CHROMATIC NUMBER. Given a graph G and an integer K. Is G K-colorable?

CLIQUE. Given a graph G and an integer K. Does G contain the complete subgraph of order K?

INDEPENDENT SET. Given a graph G and an integer K. Does G contain an independent set with cardinality K?

PLANAR SUBGRAPH. Given a graph $G = (V, E)$ and an integer K. Is there a subset $E' \subseteq E$ with $|E'| \geq K$ such that graph $G' = (V, E')$ is planar?

STEINER TREE IN GRAPH. Given a graph $G = (V, E)$, a non-negative weight for each edge of G, a subset $S \subseteq V$, and an integer B. Is there a subtree in G which contains all vertices from S and the sum of the weights of all edges of the subtree does not exceed B?

HAMILTON CYCLE. Given a graph G. Is there a Hamilton cycle in G?

For a more complete list of NP-complete problems refer to [30].

Chapter 2

Graph Planarization

2.1. Planar Graphs. Graph Planarity Criteria

A *plane* graph is a graph drawn on a plane so that no two edges of the graph (or the curves which represent them) have a common point except a vertex which is incident on them both. Thus the notion of 'plane graph' refers to the graph image rather than to the graph itself. A graph is called *planar* (*embeddable in the plane*) if it can be drawn on a plane with no two edges crossing each other. The resultant planar graph image is called an *embedding* of the graph in the plane or a *planar embedding.* Figure 2.1 shows two planar graphs, only the second of which is plane.

Along with plane embeddings of graphs we shall consider graph embeddings in other surfaces. A graph is said to be *embeddable on a surface* S if it can be drawn on S with no two edges (or, more properly, their representative curves) crossing each other.

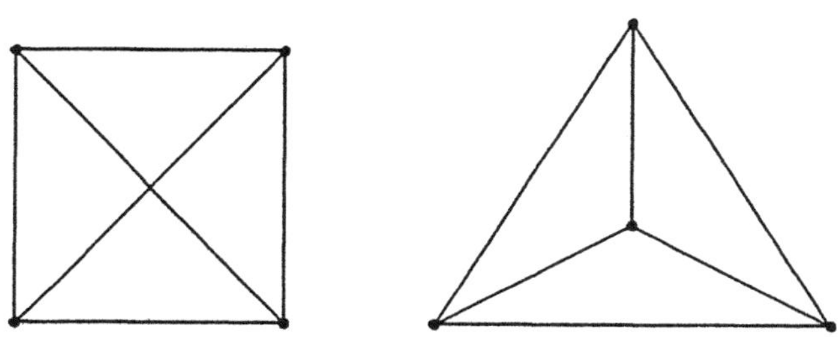

Fig. 2.1.

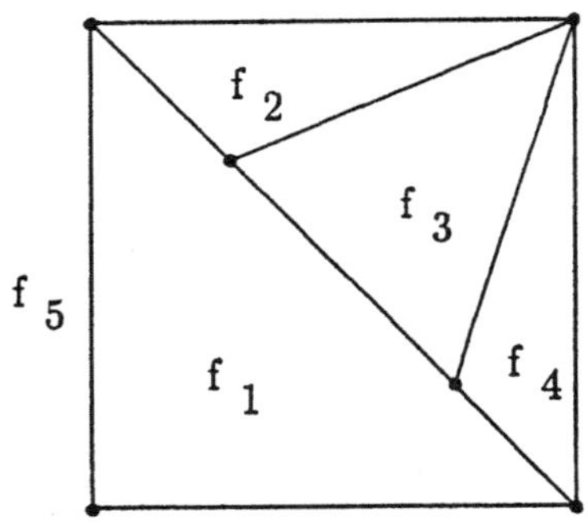

Fig. 2.2. The faces of a plane graph.

Proposition 2.1 *A graph is planar if and only if it is embeddable on a sphere.*

A *face* of a plane graph is a maximal set of points in the plane for which any two points may be joined by a curve which does not cross the edges of the graph. A face whose area is infinite is called an *external* face, and the remaining faces are considered to be *internal.* The graph of Figure 2.2, for example, has five faces and f_5 is its external face. The *boundary* of a face is the set of vertices and edges belonging to the face. It is easy to see that any internal face of a plane graph can be transformed into an external one by the so-called *stereographic projection* (Figure 2.3). For this purpose, let us draw the graph on a sphere so that no two edges cross each other (which is always possible for a planar graph). Then we shall turn the sphere so that the projection center N (i.e. the North pole) lies inside the face which we would like to make external, and then project the graph onto the plane tangent to the sphere at the South pole. In such a case the selected face becomes external.

The following theorem, which was proved by Euler in 1758, relates the number of vertices, edges and faces in a planar graph.

Theorem 2.1 *For any n-vertex connected plane graph with m edges and f faces the equality*

$$n - m + f = 2$$

holds.

An immediate corollary of this theorem is an estimate of the number of the

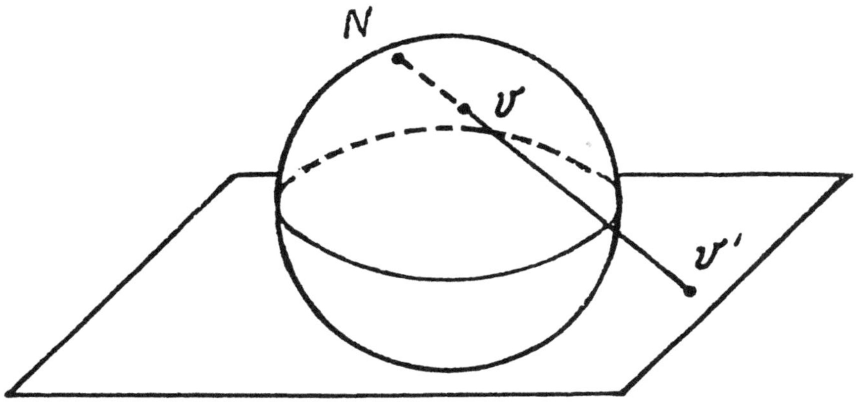

Fig. 2.3. Stereographic projection.

edges in a planar graph depending on the number of its vertices

$$m \leq 3n - 6.$$

Moreover, if graph G is biconnected and contains no triangles, then

$$m \leq 2n - 4.$$

Clearly, a graph is planar if and only if every connected component of the graph is planar. Whitney [89] has demonstrated that a graph is planar if and only if every block of the graph is planar. Thus it will suffice to restrict ourselves to biconnected graphs when studying planar graphs.

Any planar graph can be embedded in a plane in many ways. For example, there exist different ways of embedding a wheel W_5 in a plane (Figure 2.4) but its embedding on a sphere is unique. The same is true for all 3-connected graphs.

Theorem 2.2 *(Whitney [90]) Any 3-connected planar graph can be embedded on a sphere uniquely.*

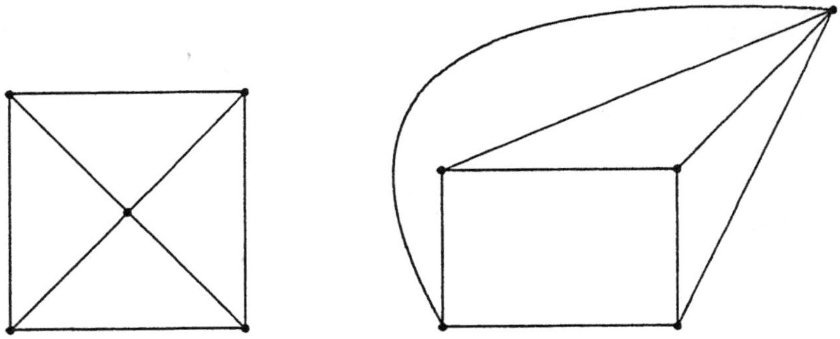

Fig. 2.4.

Specifically, the theorem implies that for any plane embedding of a 3-connected graph there exists an unique collection of face boundaries.

There exist multiple criteria of graph planarity. We shall consider characterizations of planar graphs which have been given by K. Kuratowski, H. Wagner, F. Harary and W.T. Tutte. It is worth noting that practical testing of conditions contained in these characterizations is rarely easy.

To formulate a well-known criterion of Kuratowski, we shall introduce the notion of ***homeomorphism*** of a graph. This involves ***subdivision*** of an edge $e[v_1, v_2]$ in a graph. As a result the graph has a new vertex u added and an original edge e is replaced with two new edges $e_1 = [v_1, u]$ and $e_2 = [u, v_2]$. Two graphs are said to be ***homeomorphic*** if both of them can be obtained by subdividing the edges from the same graph.

Clearly, a planar graph may be homeomorphic to the planar graph alone and all subgraphs of a planar graph are also planar. This suggests that there must exist nonplanar graphs which are minimal in the sense that all their subgraphs are planar and the degrees of all their vertices are greater than two. Examples of graphs which are minimal in the above sense are numerous. Among them are clearly K_5 and $K_{3,3}$. The criterion of Kuratowski shows that no other minimal nonplanar graphs exist, in effect.

Theorem 2.3 *(Kuratowski) A graph is planar if, and only if, it contains no subgraphs homeomorphic to* K_5 *or* $K_{3,3}$.

Let us introduce an operation of contraction of an edge $[u, v]$ in a graph. This operation results in replacing vertices u and v with a new vertex which can be denoted by u as before in the graph. The edge $[u, v]$ is removed from the graph and the $[v, v^1]$-form edges are replaced with the edges $[u, v^1]$.

Theorem 2.4 *(K. Wagner [84], F. Harary, W. Tutte [33]) A graph is planar if and only if it contains no subgraph contractable to* $K_{3,3}$ *or* K_5.

There are some other graph planarity criteria but we shall not dwell on them.

2.2. Graph Planarity Testing Algorithms

Graph planarity criteria that have been discussed above do not allow, first, to develop an efficient algorithm for graph planarity testing and, second, provide no information about planar embedding of a graph. Although the methods developed for graph planarity testing cannot be translated into an elegant criterion, they allow one to efficiently determining if a graph is planar.

Among them we can highlight a group of algorithms based on 3-connectivity [9, 42]. Unfortunately, the efficiency of these algorithms is not high. Specifically, the efficiency of the fastest of them [42] has been estimated as $O(n^3)$ though the experimentally obtained data suggest that their efficiency is somewhat better. A limitation of these methods is the fact that they do not construct a planar embedding of a graph.

The best results were derived from the design of the so called iterative algorithms for topological analysis based on the gradual graph embedding in the plane [1, 37, 62, 70]. If the procedure of graph embedding can be brought up to completion we conclude that the initial graph is planar, otherwise it is nonplanar. Among the advantages of these algorithms are high efficiency, relatively low requirement for computer memory, and the possibility of obtaining a planar embedding at graph testing. Below we shall consider this class of algorithms in more detail.

An algorithm of this type has been pioneered by Auslander and Parter in [1]. The basic idea of the algorithm consists in reducing the original problem to a collection of smaller sized problems. By this algorithm a graph is first searched for an arbitrary cycle C. After deletion of cycle C the graph is decomposed into some connected portions. The planar embedding construction algorithm is repeatedly applied to each portion merged with C. Then whenever possible the planar embeddings of individial portions are merged to form the embedding of the entire graph.

However [1] contains some inaccuracies that may lead to looping. An accurate formulation of the algorithm has been given by Goldstein [31] who described an implementation of the technique with time complexity $O(n^3)$. Then the idea of Auslander and Parter was developed in [5, 28, 37].

A most efficient method which implements and improves Goldstein's algorithm is the algorithm proposed by Hopcroft and Tarjan. Initially, by applying the one-side branching strategy, they obtained an algorithm that runs in $O(nlogn)$ time. Then the algorithm was improved to run in $O(n)$ time [37]. The basic idea behind it is to find a cycle C in a graph G, embed it in a plane as a simple closed curve, decompose the remaining portion $G - C$ into edge-nonintersecting paths, and then attempt to embed each path either entirely inside or outside C. If we succeed in embedding the whole graph in such way then it is planar, otherwise the graph is nonplanar. A major problem of the technique is the requirement that either the interior or the exterior of C may be selected for path placing, and that improper selection of placement area at an earlier stage does not prevent the placing of successive paths. Otherwise this would lead to a false conclusion that the planar graph is nonplanar. Below we discuss two iterative algorithms.

The first one based on the proposal of Demoucron, Malgrange and Pertuiset [17] applies sequential extension if a planar embedding of some initial subgraph, which could be, for instance, an arbitrary cycle, to a complete embedding of the specified graph, or states that its construction is impossible, which by selected embedding rules is equivalent to nonplanarity of the initial graph.

Another method proposed by Lempel, Even and Cederbaum [48] consists in constructing an 'extended' planar graph which can be transformed into an embedding graph by 'natural' operations. If such a transformation turns out to be impossible the original graph is considered nonplanar.

We now turn to more detailed discussion of these methods. For this purpose we introduce some definitions. Let there exist an embedding of a subgraph $\tilde{G} = (\tilde{V}, \tilde{E})$ of $G = (V, E)$. A *segment S with respect to $\tilde{G}$* is considered to be a subgraph of G having one of the following two forms:

1. edge $e = [u, v] \in E$ such that $u, v \in \tilde{V}$ and $e \notin \tilde{E}$;
2. connected components of graph $G \backslash \tilde{G}$ complemented with all their incident edges in G and their endpoints.

A vertex $v \in \tilde{V}$ in the segment S is said to be *contact vertex with respect to $\tilde{G}$*. Figure 2.5 shows the graph G with its embedded subgraph $\tilde{G}$ and all segments with respect to $\tilde{G}$. The contact vertices are encircled.

Clearly, if an embedding of G can be extended to a planar embedding of

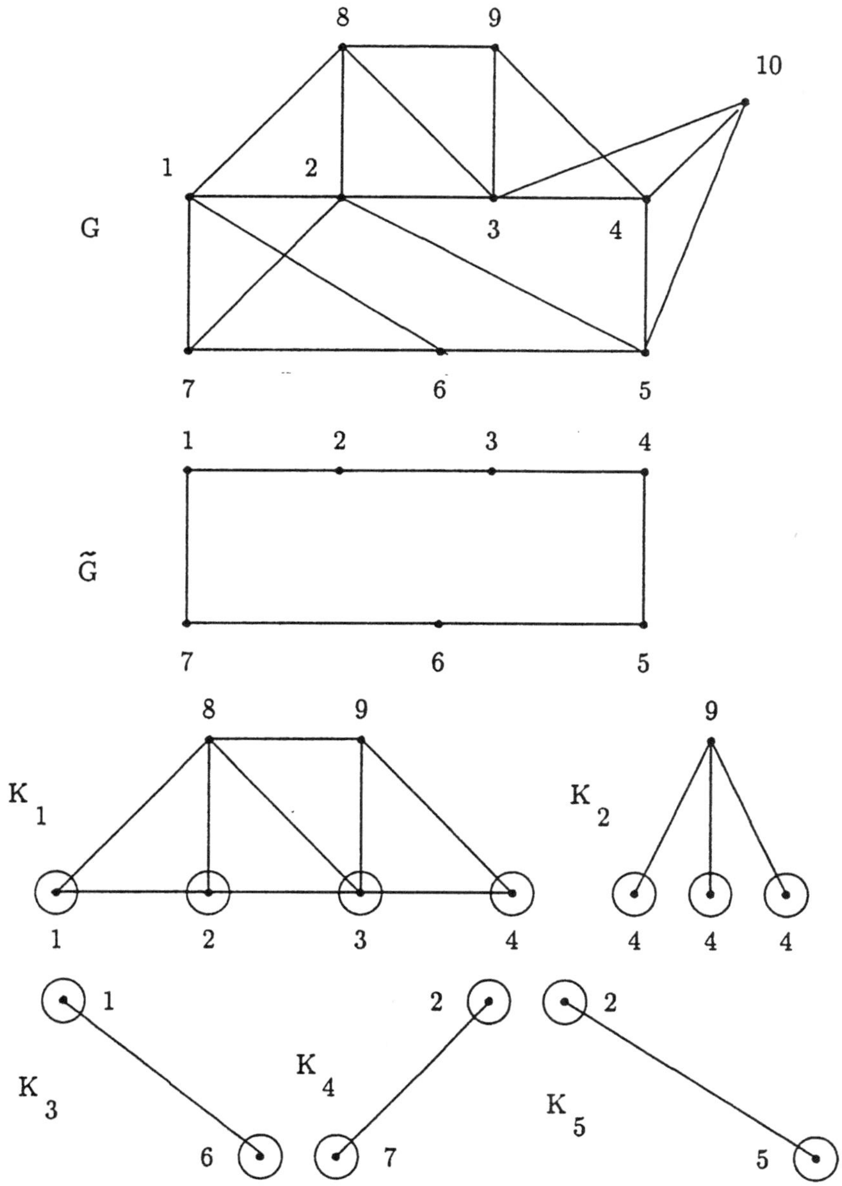

Fig. 2.5. The graph G, its embedded subgraph $\tilde{G}$, and all segments with respect to $\tilde{G}$.

G then for each segment S there exists a face of the embedding of G which contains all the contact vertices of this segment. Such a face is called *pertinent* and the set of all pertinent faces in S is denoted by $\Gamma(S)$.

The following procedure of a gradual embedding construction is naturally suggested. As an initial subgraph $\tilde{G}$ we take an arbitrary cycle of G. Then at each step we test nonemptiness of the set $\Gamma(S)$ for all segments S with respect to $\tilde{G}$. If the condition is not satisfied the planar embedding $\tilde{G}$ cannot be extended to a planar embedding of G. Otherwise we select the segment S, its pertinent face and a path P in S which connects two contact vertices. Then we embed the path P in the selected face, i.e. extend the subgraph of G, etc. The process is repeated until either an embedding of the initial graph G is constructed or we state that it cannot be constructed by these rules.

Clearly, if by applying the algorithm a planar embedding of graph G has been constructed then G is planar. It remains to be seen whether there is a case when the planar embedding cannot be constructed for a planar G by the procedure. Note that at executing the above procedure an alternative can arise in any of the following four cases:

- selection of an initial cycle;
- selection of a segment;
- selection of a pertinent face;
- selection of an embeddable path in the segment.

The basic result of [17] is the statement that planarity testing of a biconnected graph is independent of the choice of an initial cycle, a pertinent face and an embeddable path. Moreover, it will suffice to apply the following rule to make planarity testing independent of choice of segment. We select the segment with a unique pertinent face (i.e. $\mid \Gamma(S) \mid = 1$). If such a segment does not exist then we select an arbitrary segment ($\mid \Gamma(S) \mid \geq 2$ in this case, since if some segments contained an empty set of pertinent faces then nonplanarity of the initial graph would be as stated earlier).

Thus, for a biconnected graph, the above procedure allows us to find out whether the initial graph is planar and if yes to construct its planar embedding.

The procedure discussed, when applied formally, exhibits a running time $O(n^3)$, where n is the number of vertices in the initial graph. In fact, from the estimate of the number m of edges in a planar graph depending on the vertex number ($m \leq 3n - 6$) it follows that the number of steps does not exceed $O(n)$. We can apply depth-first search for segment selection and, hence, the segment determination time does not exceed $O(n)$. To determine a pertinent face it is necessary to look for faces with contact vertices. The number of such faces for each vertex does not exceed $O(n)$ and the number of searches for contact vertices equals that of edges connecting them in a segment which also does not exceed $O(n)$. Thus the time for finding the pertinent faces for all segments does not exceed $O(n^2)$. Hence the total estimate of the time complexity of the algorithm is $O(n^3)$.

It should be noted that in order to develop an efficient planarity testing algorithm based on the above procedure there is no need of computing all data on segments and their pertinent faces at each step of planar embedding construction. It will suffice to take data obtained at the earlier steps and update them. By applying these considerations Rubin [70] has developed an efficient algorithm for graph planarity testing with time complexity $O(n)$. The Karzanov algorithm [41] (refer to [82] for its implementation) will probably be competitive in running time with that of Rubin [70]. For this algorithm, the running time estimate is also linear in the number of vertices.

In 1967 Lempel, Even and Cederbaum [48] proposed another algorithm for graph planarity testing. The central idea of the algorithm is as follows.

Let $G = (V, E)$ be a biconnected graph. We shall number its vertices as $1, ..., n$ such that vertices *1* and n be adjacent, and for any other vertex v_j there exist two vertices v_i and v_k adjacent to it for which $i < j < k$. Let $s = 1, t = n$. Such numbering is called an *st-numbering*. As it has been shown in [23], any biconnected graph can be *st*-numbered. Figure 2.6 shows an example of a biconnected graph with the *st*-numbering.

We shall denote by $G_k = (V_k, E_k)$ a subgraph of a graph E induced by the vertex set $\{1, ..., k\}$. If $k < n$ then there exist the edges in G which connect a vertex of V_k with a vertex of $V \backslash V_k$. These edges are called *virtual* edges. For each virtual edge $e = [i, j]$, $i \leq k$, we shall adjoin a new virtual vertex u_e and an edge $[i, u_e]$ to G. The vertex u_e is labeled j. The graph thus constructed we shall denote by G'_k. It can be readily shown [48] that if G is planar then for any $k = 1, ..., n$ there exists a planar embedding B_k of the graph G'_k such that in B_k all the virtual edges appear in an external face (Figure 2.7).

We shall now orient the graph G'_k such that any of its arcs emanates from a vertex with a lower number and enters a vertex with a greater number. The obtained digraph we denote by D_k. Let $W(v)$ be the vertex set of D reachable from v(without the proper vertex v), $W_i(v)$, $i = 1, ..., p_v$ be the vertex set of the i-th connected component of subgraph of G'_k induced by the vertex set $W(v)$, and $T_i(v)$, $i = 1, ..., p_v$ be the subgraph of G'_k induced by the vertex set $W_i(v) \cup \{v\}$. Let v be an articulation vertex of G'_k. We shall consider two operations on the embedding B_k:

– reversing some subgraph $T_i(v)$, $1 \leq i \leq p_v$;

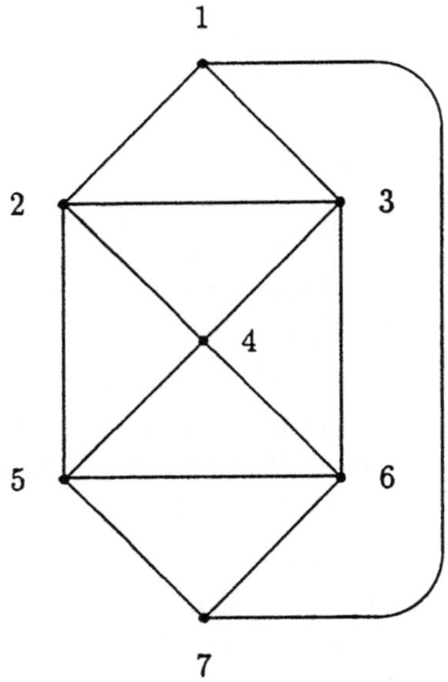

Fig. 2.6. A graph G with specified st-numbering.

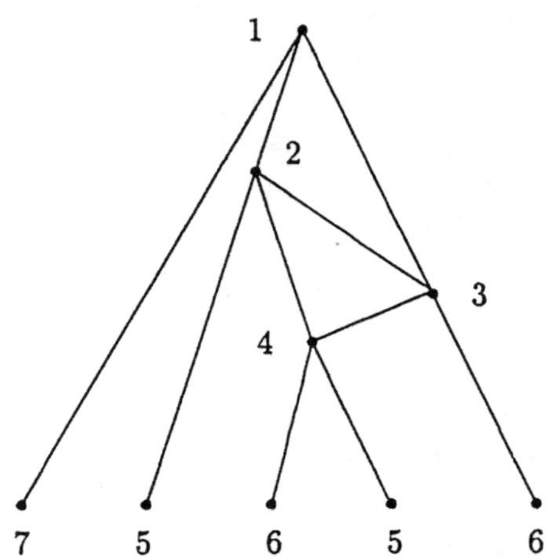

Fig. 2.7. The embedding B_4 for the graph of Figure 2.6.

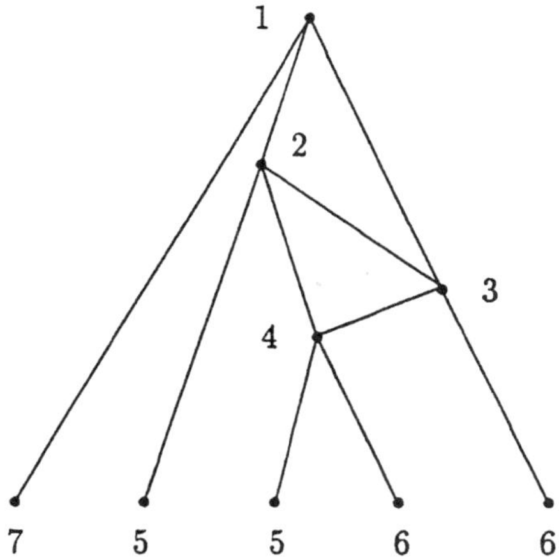

Fig. 2.8. The embedding B_4'.

– permuting subgraphs $T_i(v)$ and $T_j(v)$, $1 \le i, j \le p_v$.

Clearly, these operations do not affect the above-mentioned properties of the embedding B_k. Moreover, as [48] shows, if the graph G is planar then by using these operations we can go from any embedding B_k to an embedding B_k' of G_k' such that all vertices labeled $k+1$, $k = 1, ..., n-1$, appear consecutively (Figure 2.8).

By merging all these vertices into a single vertex $k+1$ we obtain an embedding B_{k+1} of graph G_{k+1}'. The process is repeated until either a planar embedding of G is obtained or G is proved to be nonplanar (there is then no embedding B_k' of G_k' for some k).

We can demonstrate that by applying the above planarization rules it is possible to develop an algorithm whose running time does not exceed $O(n^3)$. To develop a linear algorithm for graph planarity testing based on this method, Booth and Lueker [8] proposed using a data structure called a *PQ-tree* for representing an embedding B_k.

The PQ-tree T_k corresponding to an embedding B_k consists of three types of vertices: leaves, P-vertices and Q-vertices. Each vertex of the tree T_k is labeled.

Leaves of T_k correspond to virtual vertices of B_k and have the same labels as the edges in B_k incident with the corresponding virtual vertices.

P-vertices of T_k represent articulation vertices in G_k' and have the same

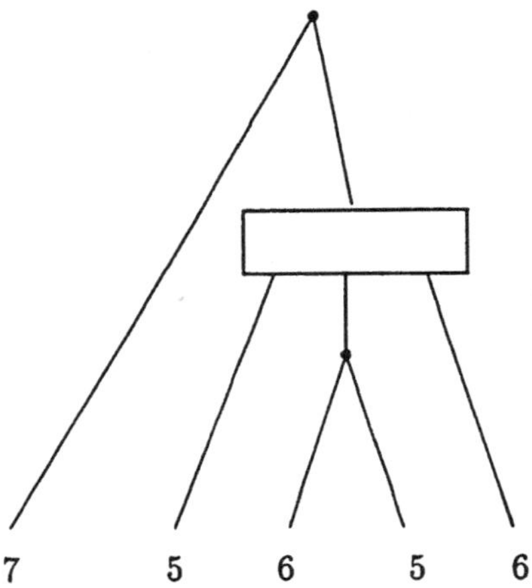

Fig. 2.9. The PQ-tree corresponding to the embedding B_4

labels as their corresponding vertices in embedding B_k. The leaf in T_k labeled as the edge $[i, j]$, $i < j$, is a child of the P-vertex corresponding to i.

Q-vertices of T_k represent the blocks of G'_k. Let $y_0, y_1, ..., y_p$ be the articulation vertices of G'_k which belong to a block of G'_k with $y_0 < y_i$ for all $i=1,...,l$. Then this block is represented by a Q-vertex whose children are the P-vertices corresponding to vertices $y_1, ..., y_p$. Moreover, the order of these P-vertices corresponds to that of vertices $y_1, ..., y_p$ in the embedding B_k. The Q-vertex is a child of the P-vertex which represents y_0.

The PQ-tree corresponding to the embedding B_k of Figure 2.7 is shown in Figure 2.9. Here the P-vertices are represented as circles and the Q-vertices as rectangles.

Two PQ-trees are said to be *equivalent* if one can be obtained from the other by performing the following two operations:

- reversing the order of the children of a Q-vertex;
- permuting the children of a P-vertex.

It can be shown that the embedding B'_k exists if and only if there exists an PQ-tree T'_k equivalent to T_k such that all the leaves labeled as a $[i, k+1]$-type edge appear consecutively. This was the basic of a linear algorithm for graph planarity testing. For more details on that algorithm refer to [13, 38].

2.3. Approximation Algorithm for Graph Planarization

If a graph planarity testing problem is solved efficiently and, moreover, linearly in the number of vertices, then the problem of nonplanar graph planarization, i.e. the removal of the minimum number of edges to make the resulting graph planar, becomes an NP-complete problem [30].

In the present section we shall discuss the approximation algorithms for solving graph planarization problems based on the Demoucron-Malgrange-Pertuiset and Lempel-Even-Cederbaum schemes discussed earlier. However, prior to starting the discussion of them let us recall that an important characteristic of any approximation algorithm A is the number ρ_A showing the worst-case relation between the solution obtained and the optimum one. Thus ρ_A is determined as a supremum of the ratio of the optimum solution to the approximate solution produced by this algorithm, i.e.

$$\rho_A = \sup \frac{t_{opt}}{t_A}.$$

In the case discussed, t_{opt} is the number of edges in a maximum planar subgraph of the original graph, and t_A is the number of edges in a planar subgraph which has been obtained as a result of executing specific approximation algorithm A.

For a wide class of NP-complete problems it is known that if the hypothesis $NP \neq P$ holds then $\rho_A = \infty$ for any polynomial approximation algorithm. Among such problems, for example, is finding a maximum complete subgraph in a graph. On the other hand, for some NP-complete problems there are well-known polynomial approximation algorithms for which precise estimates obtained exceed ρ_A. In fact, since the number of edges in a planar graph does not exceed $3n - 6$, where n is the number of vertices and the number of edges in any spanning tree is $n - 1$, then

$$\rho_A \leq 3$$

for any algorithm A which constructs a connected spanning subgraph while solving the nonplanar graph planarization problem.

A major advantage of iterative algorithms is the possibility of constructing approximation algorithms for graph planarization on their base, i.e. construction of a maximum planar subgraph which is of practical importance. All these algorithms are based on the following idea. If a required embedding cannot be constructed by sequential iteration, the obstructive edges are removed from the graph and the procedure continues. The result is a plane embedding of a subgraph of the original graph which is treated as an approximate solution of the planarization problem.

Let us consider, for instance, the Demoucron-Malgrange-Pertuiset algorithm (item 2.2) and suppose that a successive step revealed that a segment S is not embeddable, i.e. there is no face containing all contact vertices of this segment. Then removal of a ('minimal') edge set will result in partitioning of S into several segments $S_1, ..., S_k$, each of which is embeddable, and continue the process. As a result we shall obtain a plane embedding of some subgraph of the original graph which can be treated as an approximate solution of the planarization problem.

Let us now identify the basic procedures of the algorithm whose modification may cause a search among the embeddings of the planar subgraph. These procedures select:

1. the original cycle;
2. the face for embedding a segment which allows multiple embedding;
3. the embeddable segment;
4. the path in an embeddable segment;
5. the cuts in the segments for which the set of faces acceptable for their embedding is empty.

We will now look at each of the above procedures in more detail.

Generally, the number of cycles in a graph depends exponentially on the number of its vertices. Therefore searching among all cycles during execution of the proposed algorithm for construction of the maximum planar subgraph is unrealistic. We found it reasonable to select the cycles that satisfy some conditions. For instance, a maximum length or maximum weight condition (when the original graph is weighted). However, the problem of finding a maximum length cycle is NP-complete [30]. We may use various approximation algorithms for its solution.

The above considerations about original cycle selection can be extended to the path selection problem in an embeddable segment. It should be noted that when an original cycle is selected only once, path selection is performed repeatedly, hence the path selection algorithm must be efficient enough.

Consider now the face selection problem for a segment that permits a different embedding.

Let S' and S'' be two segments and f be a face pertinent for embedding segments S' and S''. Segments S' and S'' are said to be *conflicting* if there exist paths P' and P'' in segments S' and S'', respectively, such that they cross at embedding in the face f.

When selecting a face for segment embedding a problem of minimizing the conflicting segments arises during execution of the algorithm. For simplicity assume that only the initial cycle has been embedded. In that case each segment can be embedded either in the external face or the internal face. Let us determine graph the K of conflicts between segments. Two vertices are adjacent if their associated segments conflict. Then the problem of minimizing pairs of conflicting segments, which select a face for their embedding, is reduced to the problem of finding the maximum bipartite subgraph of graph K. Then the first part of this subgraph will contain the segments embeddable in the external face and the second in the internal face. This problem also belongs to the NP-complete class [30] and calls for approximation methods for its solution.

We now turn to the face selection algorithm to embed segments which allow different embedding. Suppose that the set of all pertinent faces is nonempty for each segment. The set of all segments can be divided into two subsets X and Y. X consists of segments for which the embedding face has already been determined. Initially, subset X consists of segments with a unique embedding. Then to embed each segment in Y we select a face f where the number of conflicts between the f-embeddable components from X is a minimum.

Let $X(f)$ be the set of segments to be embedded in face f. A component of $X(f)$ to be embedded in f should be selected from among the largest independent subset of vertices in the conflict graph K. Thus selection of an embeddable segment gives rise to an NP- complete problem of determining the largest independent set of vertices of the graph. At present there are many well-known approximation algorithms for finding the largest independent set. We believe it advisable to take the most time-efficient algorithms since this problem is to be solved repeatedly to determine the maximum planar subgraph.

Suppose no pertinent faces exist for segment S. Then the set C of contact vertices of S can be represented as the union

$$C = C_1 \cup \ldots \cup C_k, \tag{2.1}$$

where C_i is the set of contact vertices of segment S which belongs to face f_i. It is desired to transform the union into a partition by redistributing the vertices from subsets $C_i \cap C_j$. Such a redistribution can be performed in the same way as at selection of a face for segment embedding with the minimum number of conflicts.

Suppose that (2.1) is a partition, i.e. the subsets C_i are not pairwise intersecting. Then we shall cut the segment S so that the vertices of the sets $C_1, \ldots, C_k$ are isolated from each other. To perform the operation we can apply the well-known flow algorithms [60].

On the basic of the Lempel-Even-Cederbaum algorithm Ozawa and Takahash [59] proposed another algorithm for graph planarization. We shall consider the basic ideas of the proposed approach.

Let G be a biconnected graph whose vertices are numbered from *1* to n by using an *st*-numbering, and $T_1, ..., T_{n-1}$ be the corresponding sequence of PQ-trees. Let S_{k+1} denote the set of leaves in tree T_k labeled by edges of the form $[i, k+1]$.

The pertinent subtree of T_k is the subtree of minimum height that contains all vertices of S_{k+1} The root of the subtree is called the *pertinent root*. For an arbitrary vertex v of tree T_k, we shall find the *frontier* $F(v)$ as the sequence of all the pertinent leaves in T_k with respect to v whose order coincides with their arrangement from left to right.

We shall consider the following four types of vertices in tree T_k.

Type W is for vertices whose frontier contains no vertices from S_{k+1}.

Type B is for vertices whose frontier consists of only vertices from S_{k+1}.

Let $T_k(v)$ be a subtree of T_k induced by the vertices reachable from v and by v itself. The vertex v is said to be *Type H* if its frontier consists both of vertices from S_{k+1} and of vertices not belonging to S_{k+1} and there exists a tree equivalent to $T_k(v)$ such that all leaves of S_{k+1} appear consecutively at either the left or the right end of the frontier. A vertex is said to be *Type A* if its frontier consists both of vertices from S_{k+1} and of the vertices not belonging to S_{k+1} and there exists a tree equivalent to $T_k(v)$ such that all the leaves of S_{k-1} appear 'consecutively in the middle' of the frontier with at least one vertex not belonging to S_{k+1} appearing both at the left and the right end of the frontier.

A PQ-tree is called *reducible* if its pertinent root is Type B, H or A, otherwise it is *irreducible*. We can show that a graph G is planar if and only if all the trees T_i are reducible. It is easy to see that the trees T_1 and T_{n-1} are always reducible. An irreducible tree T_i can be made reducible be deleting some leaves. It is natural that the number of vertices to be deleted should be a minimum. By making all the trees T_i reducible in such a way, we can obtain a planar subgraph of the graph G deleting the edges associated with the deleted leaves from G.

Let T_i be an irreducible PQ-tree corresponding to an n-vertex nonplanar graph. For an arbitrary vertex v of T_i, we shall denote by w, b, h and a the minimum number of leaves reachable from v which should be deleted from T_i to make the vertex v Type W, B, H or A, respectively. A vertex's type is completely determined by the type of its children. Moreover, the w, b, h and

a number of the vertex v can be computed from that of its children.

Thus to make T_i reducible we first traverse it bottom-up from the leaves to the root and compute the value of w, b, h and a for every vertex in T_i. Once the values w, b, h and a of the pertinent root are computed, we make the root Type B, H or A, depending on which one of the numbers b, h or a is the smallest. Then by traversing top-down from the pertinent root to the leaves we make each vertex of the pertinent tree Type W, B, H or A depending on the parent type.

The obtained information makes it possible to decide which leaves should be deleted from T_i to make it reducible. The formulae for computing w, b, h and a can be found in [59].

Note that this algorithm allows deletion both of vertices in S_{i+1} and the vertices not belonging to S_{i-1}. As a result this algorithm does not necessarily allow us to determine a spanning subgraph of the specified graph.

R. Jayakumar, K. Thulasiraman and M.N.S. Swamy [39] modified Ozawa and Takahash's algorithm so that deletion of vertices not belonging to S_{i+1} from tree T_i is forbidden. As a result they obtained the $O(n^2)$-complexity algorithm with the required memory size $O(n + m)$ for constructing a spanning planar subgraph of a nonplanar graph with n vertices and m edges. On the bases of this algorithm they proposed [39] an algorithm of the same time complexity for constructing a maximum planar subgraph of the specified nonplanar graph.

For Hopcroft and Tarjan's algorihm an approximation algorithm for planarization was proposed in [17].

Chapter 3

Hypergraph Planarization

3.1. The Concept of Planar Hypergraph

A.A. Zykov was the first person to extend the notion of planarity to hypergraphs [92]. According to [92] a hypergraph is called *planar* if it has a planar Konig representation. In general, such a definition of a planar hypergraph is not adequate for the planarity of an electrical circuit. In fact, in the layout of a planar circuit the nets correspond to trees which cannot be replaced by star-type subgraphs without creating nonplanarity in all cases. The limitation of using the Zykov hypergraph planarity definition to investigate circuit planarity was shown in [6, 22, 78]. These researches suggested different mathematical models for circuits and introduced corresponding definitions of circuit model planarity which are more suitable for solving the circuit planarization problem. The necessity of solving circuit planarization problems gave rise to a new definition of a planar hypergraph [83] based on the realization concept.

Definition 3.1 *A graph $R = (V, E)$ is said to be a realization of a hypergraph $G = (V, \mathcal{H})$ if for any edge $[u, v]$ of the graph R the vertices u and v are adjacent in the hypergraph G and the subgraph $R(H)$ of R induced by the vertex set H is connected for any $H \in \mathcal{H}$.*

Definition 3.2 *A hypergraph is said to be planar if there exists a planar realization of it.*

In what follows we shall treat hypergraph planarity in that sense.

The mathematical properties of the notion introduced were investigated and their connections with well-known combinatorial objects were found in [83]. This paper started the investigation of hypergraph planarity and its results were strengthened in different directions by many authors.

The simplest examples of nonplanar hypergraphs are Kuratovsky's graph $K_{3,3}$ and the hypergraph $K'_{3,3}$ obtained from $K_{3,3}$ by adding to one of its edges a vertex which does not belong to any other edge.

To construct nonplanar hypergraphs which are not reducible to graphs (essentually hypergraphs) is not such a simple problem. Actually, to prove nonplanarity of any hypergraph it is necessary to look over all possible realizations of the hypergraph and check out every realization for its planarity

An example of a nonplanar hypergraph which is not reducible to a graph is the hypergraph $H_{9,12,3}$. This hypergraph has the vertex set $V = \{1, 2, 3, ..., 9\}$ and the edge set $\mathcal{H} = \{\{1,2,3\}, \{4,5,6\}, \{7,8,9\}, \{1,4,7\}, \{2,5,8\}, \{3,6,9\}, \{1,5,9\}, \{2,6,7\}, \{3,4,8\}, \{1,6,8\}, \{2,4,9\}, \{3,5,7\}\}$. In this hypergraph any pair of edges contains no more than one common vertex. So in any realization of $H_{9,12,3}$ the vertices incident with the same hyperedge are connected by at least two edges. Hence the number m of edges in any realization of the hypergraph $H_{9,12,3}$ complies with the inequality

$$m \geq 2 \cdot |\mathcal{H}| = 24 > 3 \cdot |V| - 6.$$

Since for planar graphs $|E| \leq 3 \cdot |V| - 6$ (see corollary of Euler theorem 2.1), then all realization of the hypergraph $H_{9,12,3}$ are not planar.

Balanced incomplete block design is placement of n different elements in m blocks, so that every block contains exactly k different elements, every element appears exactly at r different blocks and every pair of different elements (a_i, a_j) appears exactly at λ blocks.

There are two simple equalities which connect five parameters of block design

$$mk = nr,$$

$$r(k-1) = \lambda(n-1).$$

The block design theory is one of the most developed branches of combinatorial analysis. The block design can be considered as a special case of hypergraphs possessing a large amount of symmetries. In particular, for a block design with $\lambda = 1$ it is possible to get a lower bound of the edge number in the realization of corresponding hypergraphs. The edge number in such realizations is not less than $m(k-1)$, because every pair of vertices is placed at exactly one hyperedge. Hence if the condition

$$m(k-1) \geq 3n - 6$$

holds, the hypergraph determined by block design is nonplanar.

Block design with $k = 3$ and $\lambda = 1$ is called the *Steiner triple system.* For the Steiner triple system the r and m can be expressed through n as follows

$$r = \frac{n-1}{2},$$

$$m = \frac{n(n-1)}{6}.$$

The condition $n = 1, 3$ (*mod* 6) is necessary and sufficient for the existence of a Steiner triple system. Every Steiner triple system for which this condition

$$m = \frac{n^2 - n}{6} > \frac{3n-6}{2}$$

is valid determines a nonplanar hypergraph. The smallest value of n for which this condition is valid is $n = 9$. The hypergraph $H_{9,12,3}$ described above is actually a Steiner triple system with $n = 9$, $m = 12$, $r = 4$.

Using the block design theory's results allows us to construct an infinite series of nonplanar hypergraphs and clarify the structure of nonplanar hypergraphs.

Notice that the hypergraph obtained from $H_{9,12,3}$ by removing any hyperedge is nonplanar too, because the number m^* of edges of any of its realizations complies with the condition $m^* \geq 2m \geq 3n-6$. If $m = 11$, $22 > 3 \cdot 9 - 6 = 21$, then $m^* \geq 21$.

It is easy to show that if Konig representation of a hypergraph is planar then the hypergraph is planar (according to the Definition 3.2). In fact in the Konig representation of a hypergraph all vertices of the same hyperedge $H_i \in \mathcal{H}$ are connected by a star with an additional center vertex c_i, corresponding to the hyperedge H_i. We can get rid of this additional vertex c_i by contracting in the Konig representation any edge $[c_i, x_i]$, where $x_i \in H_i$. Contracting can not turn a planar graph into a nonplanar one. Hence we can obtain a planar realization of a hypergraph with a planar Konig representation.

On the other hand there are hypergraphs which are planar according to Definition 3.2, but their Konig representations are nonplanar. The example of such a hypergraph is the hypergraph $H_{9,6,3}$ with 9 vertices, 6 hyperedges and 3 vertices in every hyperedge: $H_{9,6,3} = \{\{1,2,3,...,9\}, \{\{1,2,3\}, \{4,5,6\}, \{7,8,9\}, \{1,4,7\}, \{2,5,8\}, \{3,6,9\}\}$. The planar realization of this hypergraph is shown in the Figure 3.1. Its Konig representation is homeomorphic to the graph $K_{3,3}$.

Properly, hypergraph characteristics (which are not graph characteristics) are determined to a large extent by characteristics of the structure of all

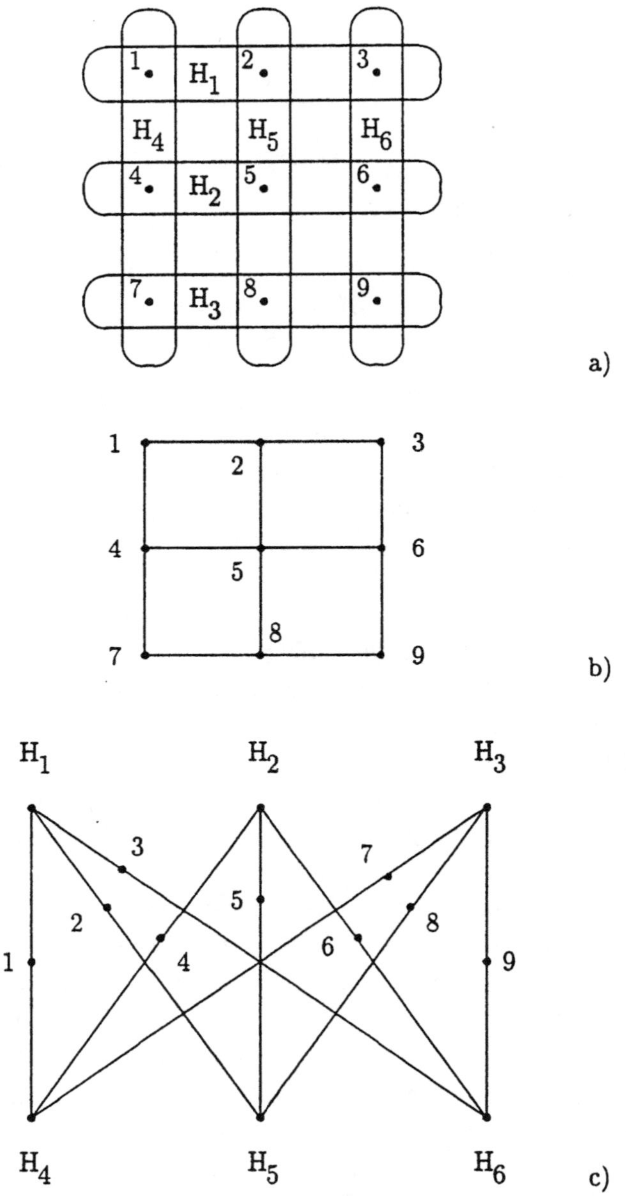

Fig. 3.1. A planar hypergraph with nonplanar Konig representation: a) the hypergraph; b) its planar realization; c) Konig representation.

hyperedge intersections. It is unexpected that hypergraph planarity does not depend on the structure of all its hyperedge intersections and depends only on family of the intersections which are minimal with respect to the partial order $\subseteq$.

Let $G = (V, \mathcal{H})$ be a finite hypergraph with a vertex set V and a family of hyperedges $\mathcal{H}$. Consider a set $\mathcal{M}$ of all minimal non-empty intersections of hyperedges from $\mathcal{H}$. Select a vertex x_M in every set $M \in \mathcal{M}$. Consider a hypergraph $G' = (V', \mathcal{H}')$ by removing all vertices other than x_M, $M \in \mathcal{M}$ from the hypergraph G.

Theorem 3.1 *[83] If the hypergraph G' is planar then the hypergraph G is planar.*

Proof. Let us associate every edge of the hypergraph G with an integer i, $1 \leq i \leq n$, $n = \mid \mathcal{H} \mid$. We index every vertex $v \in V$ of the hypergraph G by the list of the numbers of all incident with v edges of the hypergraph G. Notice, that every vertex associated with the list $(i_1, i_2, ..., i_k)$ is contained in $H_{i_1} \cap H_{i_2} \cap ... \cap H_{i_k}$. But not every vertex $v \in H_{i_1} \cap H_{i_2} \cap ... \cap H_{i_k}$ has to be associated with the list $(i_1, i_2, ..., i_k)$. Let $p = H_{i_1} \cap H_{i_2} \cap ... \cap H_{i_k}$. Denote by $\nu(p)$ the number of the vertices associated with the list $(i_1, i_2, ..., i_k)$. If $H_{i_1} \cap H_{i_2} \cap ... \cap H_{i_k}$ is a minimal (with respect to $\subseteq$) intersection, then all vertices from this intersection are associated with the same list $(i_1, i_2, ..., i_k)$. Let P be the set of all nonempty intersections p of the hypergraph G edges for which $\nu(p) > 0$, $\mathcal{M} = \mathcal{M}_0$ be the set of all its minimal elements (with respect to $\subseteq$), M_1 be the set of all minimal elements in $P \backslash \mathcal{M}_0$, $\mathcal{M}_i$ be the set of all minimal elements in $P \backslash \bigcup_{k=0}^{i-1} \mathcal{M}_k$. It is obvious that $P = \bigcup_{k=0}^{h} \mathcal{M}_k$ for some $h \geq 0$.

Let there exist a planar realization of the hypergraph G'. we will show how to get planar realization of hypergraph G.

Step 1. For every intersection $M \in \mathcal{M}_0 = \mathcal{M}$ choose one vertex from this intersection. Place the chosen vertices on the plane and connect them by edges in the same way that corresponding elements of M are connected in the planar embedding of the hypergraph G' realization. Denote the constructed graph by R_0. It is clear that in the graph R_0 any two vertices from the same hyperedge H of the hypergraph G are connected by the chain of which all vertices are incident on the hyperedge H (H-chain). Then for every vertex v of R_0 we add $\nu(M) - 1$ new vertices and connect them with the vertex v. Denote the constructed graph by R_1.

Step 2. Consider now vertices of the hypergraph G with list $(i_1, i_2, ..., i_k)$, where $H_{i_1} \cap H_{i_2} \cap ... \cap H_{i_k} \in \mathcal{M}_1$. Choose one vertex v_1 from every $M_1 \in \mathcal{M}_1$ and place it inside the face of plane graph R_1, the boundary of which contains $v \in M$, $M \subset M_1$. Add the edge $[v_1, v]$. Since $v \in M \subset M_1$ and $v_1 \in M_1$, then if the vertex v_1 and the hyperedge H are incident then v and H are incident too. Hence, if v_1 and H are incident and a vertex w placed on a plane earlier is incident with H too, then v_1 is connected with w by H-chain, because v is connected with w by H-chain. Then for every such vertex v_1 we add $\nu(M_1) - 1$ new vertices and connect them with the vertex v_1. Denote the constructed graph by R_2.

Let the graph R_i be constructed by i steps. This graph contains all vertices from all elements of sets $\mathcal{M}_0$, $\mathcal{M}_1$,..., $\mathcal{M}_{i-1}$.

Step $i+1$. Consider vertices of the hypergraph G with lists corresponding to the elements from $\mathcal{M}_i$. Choose one vertex v_i from every $M_i \in \mathcal{M}_i$. Place every such vertex inside any face of the graph R_i, which contains the vertex $v_{i-1} \in M_{i-1} \in \mathcal{M}_{i-1}$, $M_{i-1} \subset M_i$. Connect these vertices by the edge $[v_i, v_{i-1}]$. Since $v_{i-1} \in M_{i-1} \subset M_i$ and $v_i \in M_i$, then if the vertex v_i and the hyperedge H are incident v_{i-1} and H are then incident too. According to the definition of R_i the vertex v_{i-1} is connected by an H-chain with any vertex of the hyperedge H, placed on the plane so far. Consequently v_i is connected by an H-chain with any vertex placed on the plane so far, as well. Then for every such vertex v_i we add $\nu(M_i) - 1$ new vertices and connect them with the vertex v_i.

At the last $(n+2)$-th step we place on the plane the vertices of G with degree 1 (the length of the corresponding list is 1). Every such vertex can be connected with a vertex of the graph R_{n+1} if its list contains the index of the corresponding hyperedge. Denote this planar graph by R_{n+2}.

Obviously, the constructed graph R_{n+2} is a planar realization of the hypergraph G. The theorem is proved. □

Notice. The number of vertices in the hypergraph G' is not more than that of the hypergraph G. Hence clearly G' is isomorphic to G if $|\mathcal{M}| = |V|$.

Definition 3.3 *The hypergraph $G = (V, \mathcal{H})$ is said to be irreducible if G' is isomorphic to G.*

Thus, the study of hypergraph planarity can be reduced to the study of irreducible hypergraphs. For the characterization of irreducible hypergraphs a major notion is Δ-separability of the vertices in the hypergraphs.

Definition 3.4 *The two vertices u and v of the hypergraph $G = (V, \mathcal{H})$ are said to be Δ-separable if there exist two hyperedges H_1 and H_2 from $\mathcal{H}$ such that $u \in H_1$, $v \in H_2$ and $u, v \notin H_1 \cap H_2$.*

Proposition 3.1 *The vertices u and v of the hypergraph $G = (V, \mathcal{H})$ are Δ-separable if and only if the intersection of all hyperedges incident with u does not contain v, and, vice versa, the intersection of all hyperedges incident with v does not contain u.*

Proof. *Necessity.* Let the vertices u and v be Δ-separable in the hypergraph $G = (V, \mathcal{H})$, let $\{U_1, U_2, ..., U_r\}$ be the set of all hyperedges incident with u, and $\{V_1, V_2, ..., V_s\}$ be the set of all hyperedges incident with v. According to Definition 3.4 there exist a hyperedge U_i which is not incident with v and a hyperedge V_j which is not incident with u. Consequently $U_1 \cap U_2 \cap ... \cap U_r$ does not contain v, and $V_1 \cap V_2 \cap ... \cap V_s$ does not contain u. Necessity is proved.

Sufficiency. Let the intersection $U_1 \cap U_2 \cap ... \cap U_r$ of all hyperedges incident with u not contain v, and the intersection $V_1 \cap V_2 \cap ... \cap V_s$ of all hyperedges incident with v not contain u. Then there exist at least one hyperedge U_i from $\{U_1, U_2, ..., U_r\}$ which is not incident with v, and at least one hyperedge V_j from $\{V_1, V_2, ..., V_s\}$ which is not incident with u. Consequently, $u \in U_i$, $v \notin U_i$ and $v \in V_j$, $u \notin V_j$. Hence, $x, y \notin U_i \cap V_j$, i.e. the vertices u and v are Δ-separable. Sufficiency is proved. □

Corollary 3.1 *Any two vertices of the hypergraph $G = (V, \mathcal{H})$ are Δ-separable if and only if for any vertex $v \in V$ the intersection of all hyperedges incident with v is equal to $\{v\}$.*

Proof. *Necessity.* Let any two vertices of the hypergraph G are Δ-separable, i.e. any vertex $u \in V$ is Δ-separable with any vertex $v \in V \backslash \{u\}$. This means that among the hyperedges $U_1, U_2, ..., U_r$ incident with u there exists at least one hyperedge U_i which is not incident with any vertex $v \in V \backslash \{u\}$. Thus $U_1 \cap U_2 \cap ... \cap U_r = \{u\}$. Necessity is proved.

Sufficiency. Let u be an arbitrary vertex of the hypergraph G, the hyperedges $U_1, U_2, ..., U_r$ are incident with u and $U_1 \cap U_2 \cap ... \cap U_r = \{u\}$. It is clear that for every vertex $v \in V \backslash \{u\}$ there exists a hyperedge U_i which is not incident with v. On the other hand if $V_1, V_2, ..., V_s$ are hyperedges which are incident with $v \in V \backslash \{u\}$ then $V_1 \cap V_2 \cap ... \cap V_s = \{v\}$, i.e. there exists at least one hyperedge V_j which is not incident with u. Consequently, any vertex $u \in V$ is Δ-separable with any other vertex $v \in V\backslash\{u\}$. Sufficiency is proved. □

The planar hypergraph $H_{9,9,3}$ (Figure 3.1) and nonplanar $H_{9,12,3}$ described above present examples of hypergraphs with all Δ-separable vertices, as well as any hypergraph determined by Steiner's triple system.

Theorem 3.2 *[83] A hypergraph is irreducible if and only if any two vertices in the hypergraph are Δ-separable.*

Proof. *Necessity.* Let G be irreducible. Then according to the definition of the hypergraph G' any intersection of all hyperedges incident with the arbitrary vertex v of G is equal to $\{v\}$. So by Corollary 3.1 any two vertices of the hypergraph G are Δ-separable.

Sufficiency. Let any two vertices of the hypergraph G be Δ-separable. Then by Corollary 3.1 the intersection of all hyperedges incident with the arbitrary vertex u is equal to $\{u\}$. Hence, any nonempty intersection of the hyperedges is reduced to the single vertex u and consequently it is minimal with respect to $\subseteq$. Thus the set of minimal intersection of hyperedges of G is equal to the set of hyperedges of G, and the hypergraphs G and G' are isomorphic. The theorem is proved. □

An irreducible hypergraph can contain loops, i.e. hyperedges consisting of a unique vertex. With the loops removed from an irreducible hypergraph the hypergraph may become reducible. By alternatively applying reduction and loop removal we can eventually obtain a hypergraph G^* which admits no further reduction from the initial hypergraph G. The theorem implies that if G^* is planar then G is also planar. Unfortunately, however, the conditions of the theorem are not necessary; put another way, there exists a planar hypergraph G for which the hypergraph G^* is nonplanar (Figure 3.2).

Let $G = (V, \mathcal{H})$ be a finite hypergraph. Consider the semilattice $\mathcal{P}(G)$ generated by its hyperedges, i.e. the set of all nonempty intersections of the hyperedges, partially ordered with respect to inclusion $\subseteq$. We will represent semilattices by the usual graphic diagrams. For example, in Figure 3.3 the semilattice for the hypergraph $G = (V, \mathcal{H})$ is presented, where the vertex set $V = \{v_1, v_2, ..., v_8\}$ and the edge set $\mathcal{H} = \{H_1 = \{v_1, v_4\}, H_2 = \{v_2, v_3, v_4\}, H_3 = \{v_1, v_2, v_8\}, H_4 = \{v_5, v_6, v_8\}, H_5 = \{v_6, v_7, v_8\}\}$. Everywhere henceforward for convenience (as is done in Figure 3.3) we will mark every element of

G

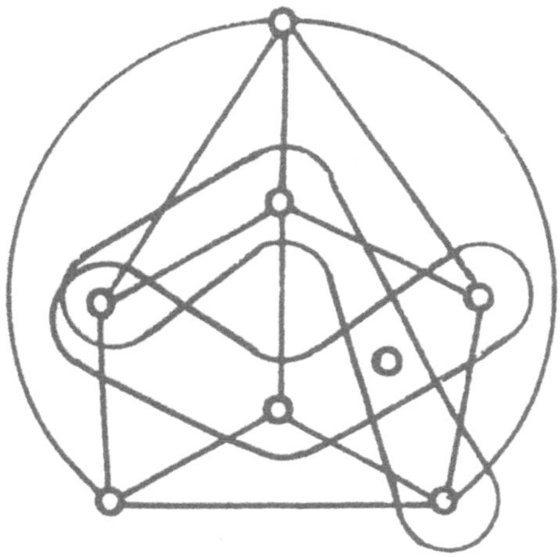

G^*

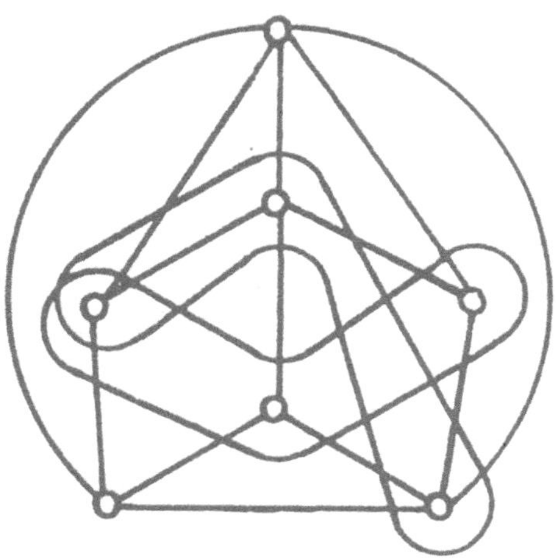

Fig. 3.2. A planar hypergraph G and nonplanar hypergraph G^*.

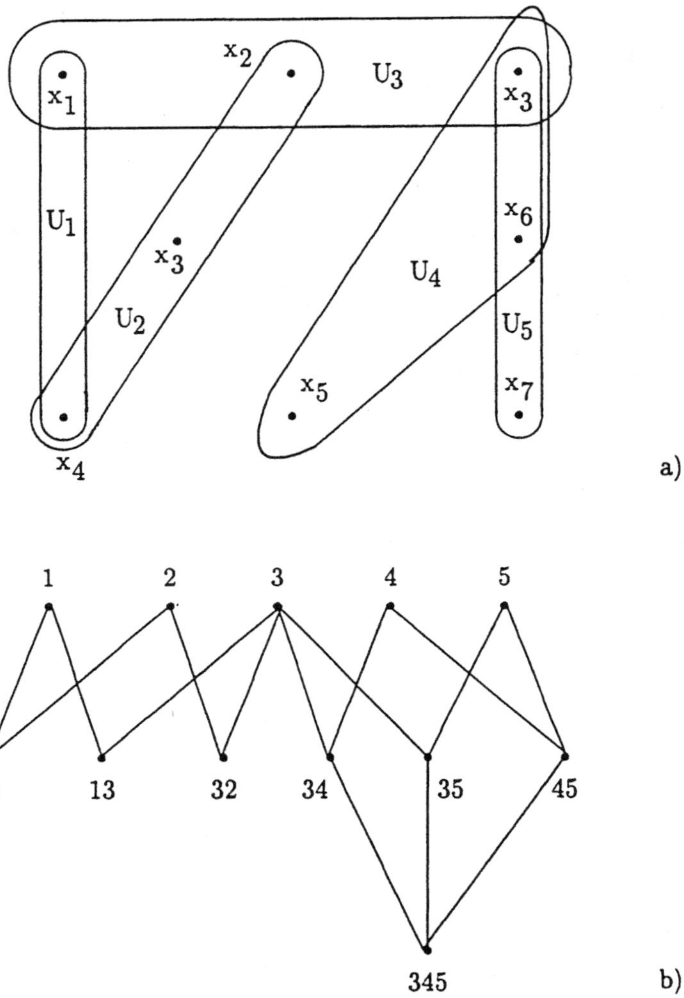

Fig. 3.3. Hypergraph (a) and semilattice (b), generated by its edges.

the semilattice (i.e. hyperedge intersection) by the list of the corresponding hyperedges. The length of this list will be called the semilattice element level.

It is easy to see that all hypergraphs with a given number of edges and with the same irreducible hypergraph have the same semilattice. Moreover, according to the definition the vertex set of an irreducible hypergraph is the set of minimal elements of the semilattice, and the hyperedge U_i contains the minimal element if the list marking this element contains the number i.

The semilattice segment is such a subset S of the elements of the semilattice $\mathcal{P}(G)$, that if $s_1 \in S$ then any element s_2, $s_1 \subseteq s_2$ belongs to S as well.

The least segment containing the given set is named the segment generated by the given set of the semilattice elements.

Let E be a graph. We will say that a semilattice $\mathcal{P}_E$ is generated by a given graph E if this semilattice is the semilattice of the hypergraph G, the line graph $L(G)$ of which is isomorphic to E, and G has the maximum number of nonempty intersections of the hyperedges among the hypergraphs with such a line graph. The example of the graph E and semilattice $\mathcal{P}_E$ generated by E is shown in Figure 3.4.

It is obvious that the semilattice of any hypergraph with a line graph E is isomorphically embeddable into $\mathcal{P}_E$.

Let us name as *p*-cutset of the semilattice $\mathcal{P}_E$ the set R of pairwise incomparable elements of $\mathcal{P}_E$, that segment, generated by the set R contains all elements of the second level of the semilattice $\mathcal{P}_E$. For example, the *p*-cutsets of the semilattice in the Figure 3.4 are the sets $R_1 = \{0,1,2,3\}$, $R_2 = \{012, 03, 13, 23\}$, $R_3 = \{012, 013, 12, 23\}$, $R_4 = \{012, 013, 023\}$.

It is easy to see that an exhaustive search of all possible *p*-cutsets of the semilattice $\mathcal{P}_E$ (and generated by *p*-cutset semilattice segments) allows us to get all possible semilattices for hypergraphs with line graph isomorphic to E.

On the other hand every *p*-cutset of the semilattice $\mathcal{P}_E$ defines an irreducible hypergraph with line graph isomorphic to E.

Notice that if two *p*-cutsets R_1 and R_2 generate the segments $S(R_1)$ and $S(R_2)$ which are isomorphic as semilattices, then the irreducible hypergraphs defined by the *p*-cutsets R_1 and R_2 are isomorphic as hypergraphs. Everything stated above allows us to organize an exhaustive search of all nonisomorphic hypergraphs with line graph isomorphic to the given one.

Theorem 3.3 *[83] All hypergraphs with no more than 6 edges are planar.*

Proof. According to Theorem 3.1 it is sufficient to prove this theorem only for irreducible hypergraphs. It will be done by an exhaustive search of all possible irreducible hypergraphs and by constructing their plane realizations.

The exhaustive search can be organized as follows. We have to sort out all graphs with number of vertices no more than 6 and then for every such graph E to sort out (as described above) all possible irreducible hypergraphs with line graph isomorphic to E.

Examples of such an exhaustive search for hypergraphs with four, five and six hyperedges are presented in Figures 3.4- 3.9. □

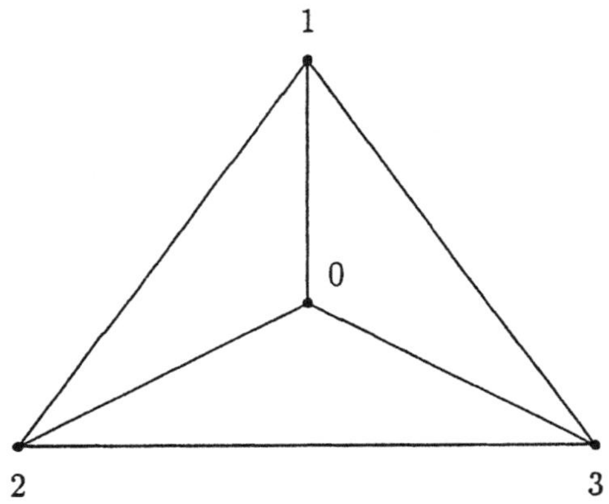

a)

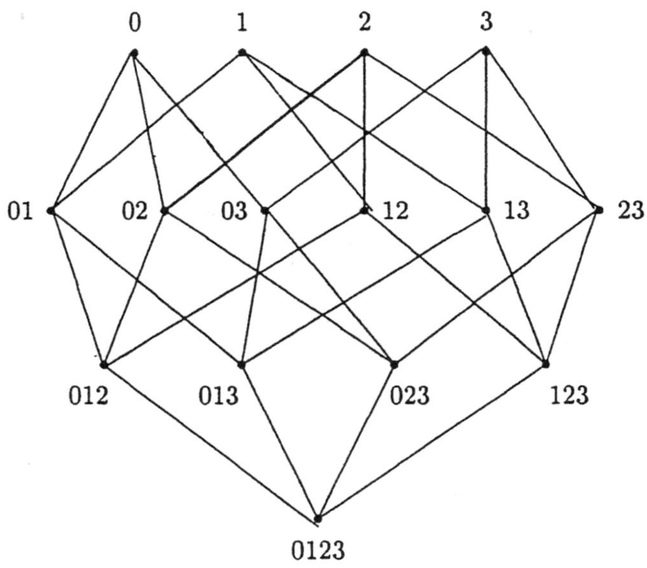

b)

Fig. 3.4. A line graph (a) of the hypergraph with four edges, and the corresponding semilattice (b).

It can be shown that if the line graph has articulation points then the corresponding irreducible hypergraph has loops. So it is enough to consider only all biconnected graphs as the line graphs of some irreducible hypergraphs.

p-cutset	plane realization of the corresponding irreducible hypergraph	list ot the p-cutsets generating isomorphic irreducible hypergraphs
01,02,03,12,13,23	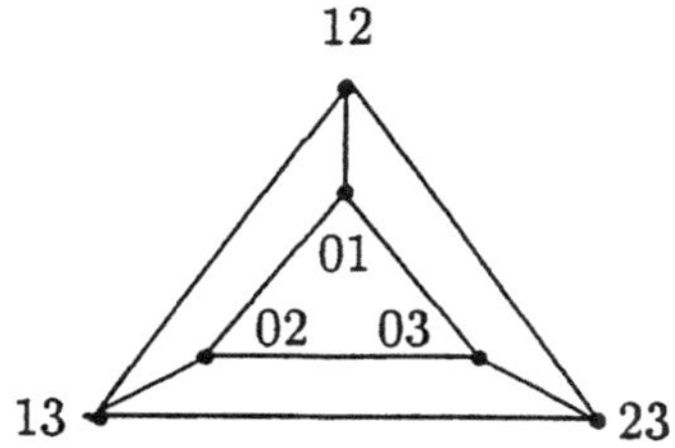	
012,03,13,23	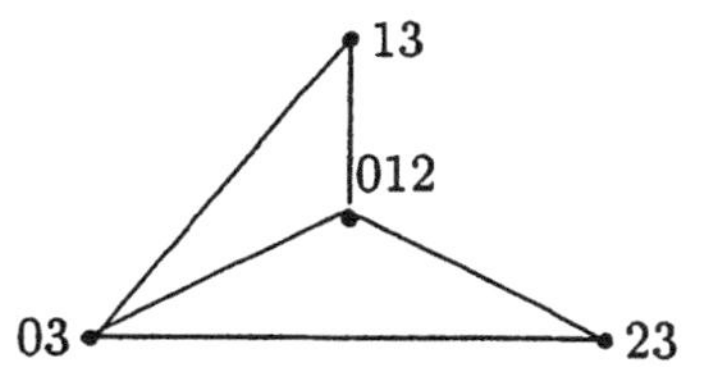	013,02,12,23 023,01,12,13 123,01,02,03
012,013,23	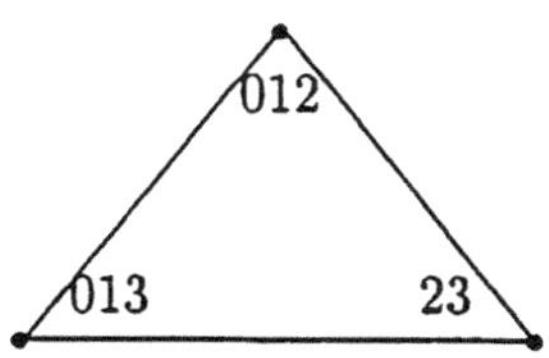	012,023,13 012,123,03 013,023,12 013,123,02 023,123,01
012,013,023	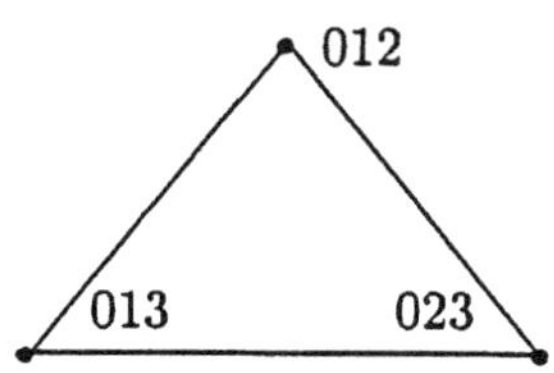	013,023,123 012,023,123
012,013,023,123	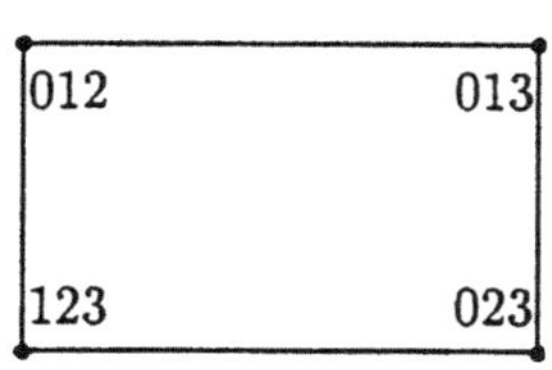	

Fig. 3.5. Exhaustive search of all irredusible hypergraphs with four edges and planar line graph from Figure 3.4.

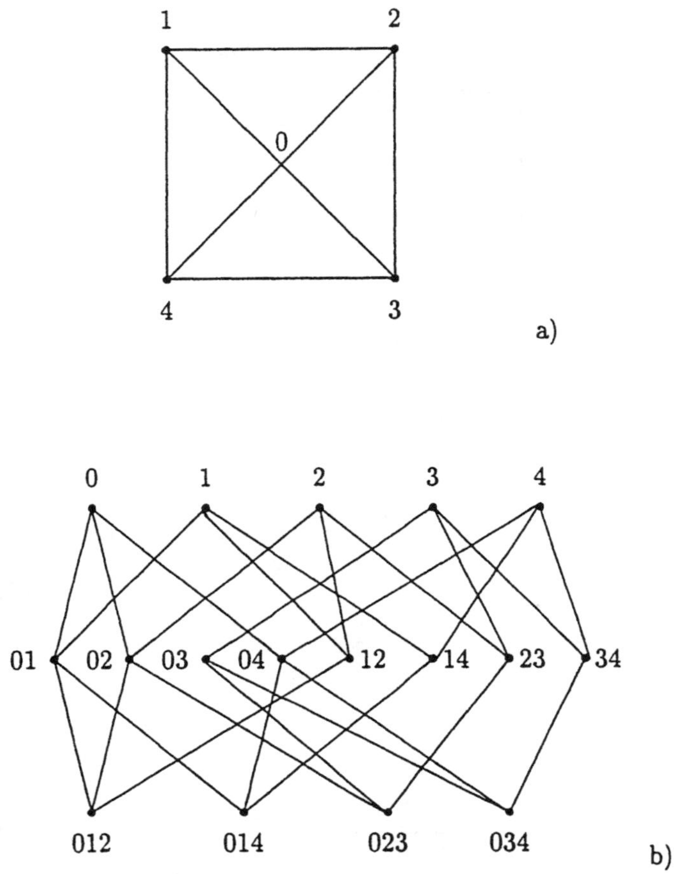

Fig. 3.6. A line graph (a) of the hypergraph with five edges and its corresponding semi-lattice (b).

Let G be a planar hypergraph with planar line graph. We will say that the plane realization R of the hypergraph G is matched to the plane embedding L of the line graph $L(G)$ if there exists a correspondence between faces of R and faces of L such that when passing around the boundary of any face f_R of the realization R the vertices of the hypergraph are arranged according to the order of the line graph vertices when passing arround the corresponding face f_L of the embedding L. In other words, if $H_1, H_2, ..., H_k$ is the boundary of the face f_L of L, then during passing the boundary of the face f_R of the realization R the vertices of the edge H_1 meet first, then follow those of H_2,

p-cutset	plane realization of the corresponding irreducible hypergraph	list ot the p-cutsets generating isomorphic irreducible hypergraphs
01,02,03,04, 12,14,23,34	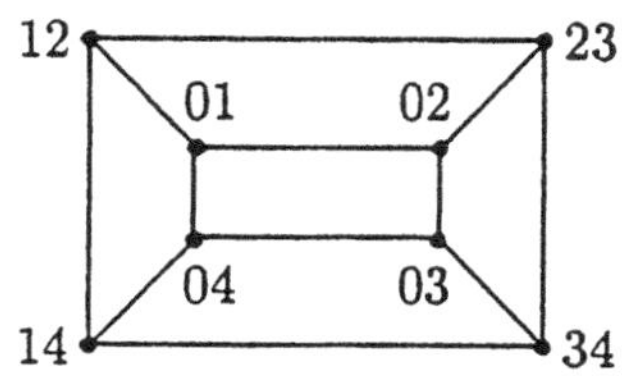	
012,03,04,14, 23,34	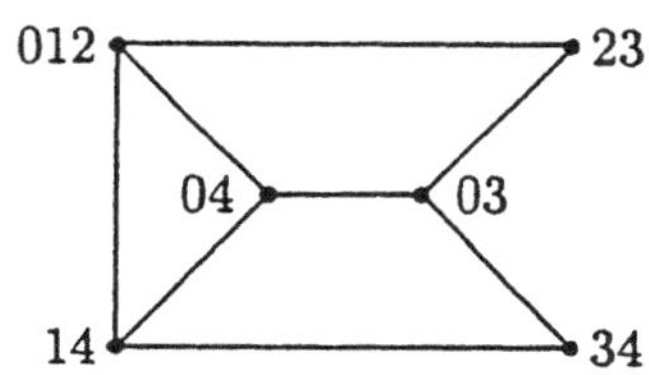	014,02,03,12,23,34 023,01,04,12,14,34 034,01,02,13,14,23
012,014,03,23,34	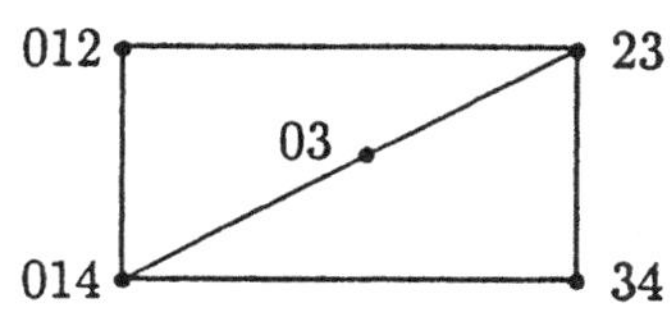	012,023,04,14,34
012,034,14,23	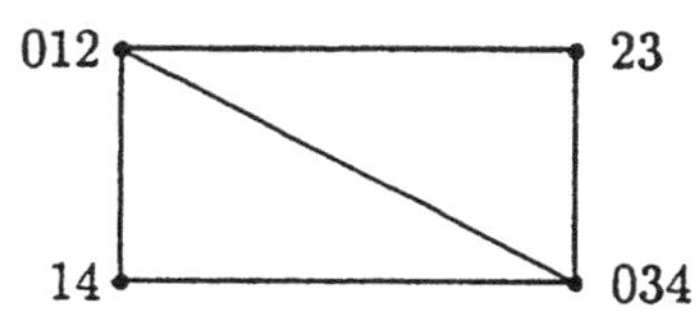	014,023,12,34
012,014,023,34		012,023,034,14 014,023,034,12
012,014,023,034		

Fig. 3.7. Exhaustive search of all irredusible hypergraphs with five edges and planar line graph from Figure 3.6.

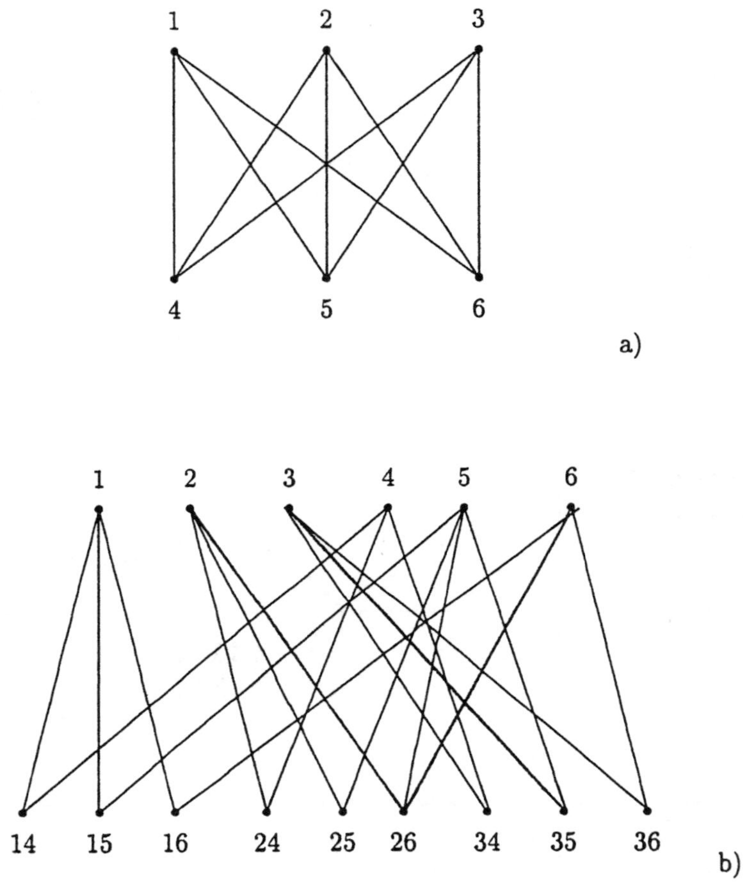

Fig. 3.8. A line graph (a) of the hypergraph with six edges and corresponding semilattice (b).

etc.

Proposition 3.2 *Let the hypergraph G have no more than 6 edges and its line graph be planar. Then for every plane embedding L of the line graph there exists a plane realization of G matching L.*

Proof. It is clear (see proof of the Theorem 3.1) that if there exists a plane realization of the hypergraph G' which is matched with the plane embedding L of the line graph of G', then the plane realization of G may be constructed easily from the plane realization of the hypergraph G'. Thus the

<table>
<tr><th>p-cutset</th><th>plane realization of the corresponding irreducible hypergraph</th><th>list ot the p-cutsets generating isomorphic irreducible hypergraphs</th></tr>
<tr><td>14,15,16,24,25, 26,34,35,36</td><td>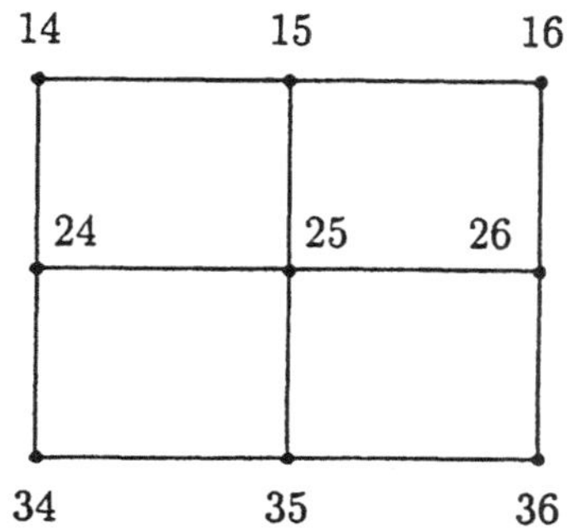
</td><td></td></tr>
</table>

Fig. 3.9. Exhaustive search of all irredusible hypergraphs with six edges and planar line graph from Figure 3.8.

exhaustive search of all irreducible hypergraphs with planar line graphs and directed construction of the plane realizations for such hypergraphs provide proof of this Proposition. Examples of the plane realizations matching the plane embeddings of the line graphs are presented in the Figures 3.4- 3.9. □

In [64] the result was strengthened for eight edges.

Theorem 3.4 *[83] The hypergraph with planar line graph is planar and for every plane embedding L of the line graph there exists a plane realization of the hypergraph matching L.*

Proof. The proof will be carried out by induction on the edge number m in the hypergraph. Here we will present only a draft of the proof. A detailed proof of this theorem can be found in [50].

For hypergraphs with no more than 6 edges the theorem is valid, according to Proposition 3.2. Let the theorem be valid for all hypergraphs with $m < k$.

Consider the hypergraph $G_k = (V, \mathcal{H})$, $|\mathcal{H}| = k$. Since the line graph $L(G_k)$ of the hypergraph G_k is planar, then $L(G_k)$ contains at least one vertex of degree no more than 5.

Let (for simplicity) the hyperedges be marked by numbers 0, 1,...,$k-1$, the vertex i of the line graph $L(G_k)$ correspond to the edge H_i, and let the vertex 0 have degree $deg(0) \leq 5$.

Let $S(0)$ be the set of the $L(G_k)$ vertices adjacent to the vertex 0. Denote by $S(H_0) \subseteq \mathcal{H}$ the set of hyperedges adjacent to H_0, including H_0 itself. Since $|S(0)| \leq 5$ then $|S(H_0)| \leq 6$.

Consider the hypergraph G_{k-1} constructed from G_k by removing the edge H_0 and removing all vertices incident with H_0. It is obvious that by removing the vertex 0 and all edges incident with it from the plane embedding L_k of $L(G_k)$ we get the plane embedding L_{k-1} of the line graph $L(G_{k-1})$ of the hypergraph G_{k-1}. It is possible that such a graph contains several 'redundant' edges connecting the vertices incident with H_0. But it does not matter for the theorem's proof.

According to the induction hypothesis there exists a plane realization of the hypergraph G_{k-1} matching L_{k-1}.

Consider a trace G_k/H_0 of the hypergraph G_k on the edge H_0. G_k/H_0 is the hypergraph with the vertex set H_0 and the edge set $\{H_i' = H_i \cap H_0 : H_0 \in S(H_0)\}$.

It is evident that the line graph of the hypergraph G_k/H_0 is isomorphically embeddible in a subgraph of the line graph $L(G_k)$, generated by the vertex set $S(0) \cup \{0\}$ (Figure 3.10), and the edge number of the hypergraph G_k/H_0 is no more than 6.

According to Proposition 3.2 the hypergraph G_k/H_0 is planar and there exists its plane realization matching the plane embedding L of the line graph $L(G_k/H_0)$ generated by the plane embedding L_k.

It can be shown that the plane realization of the hypergraph G_k matching L_k can be obtained by combining the plane realizations of the hypergraphs G_{k-1} and G_k/H_0 matching the plane embedding L_{k-1} and L respectively. □

The planar hypergraph with 18 vertices and 9 edges: $H_1 = \{1,2,10,11\}$, $H_2 = \{1,3,13,14\}$, $H_3 = \{2,3,16,17\}$, $H_4 = \{4,5,10,12\}$, $H_5 = \{4,6,13,15\}$, $H_6 = \{5,6,16,18\}$, $H_7 = \{7,8,11,12\}$, $H_8 = \{7,9,14,15\}$, $H_9 = \{8,9,17,18\}$ and the nonplanar graph $K_{3,3}$ have isomorphic line graphs. This means that the hypergraph planarity criterion can not be expressed in terms of a line graph.

The adopted definition of hypergraph planarity and the properties of planar graphs imply a series of statements that will be used in the following. Thus

1. removal and/or insertion of a vertex of degree 1 does not affect hypergraph planarity;

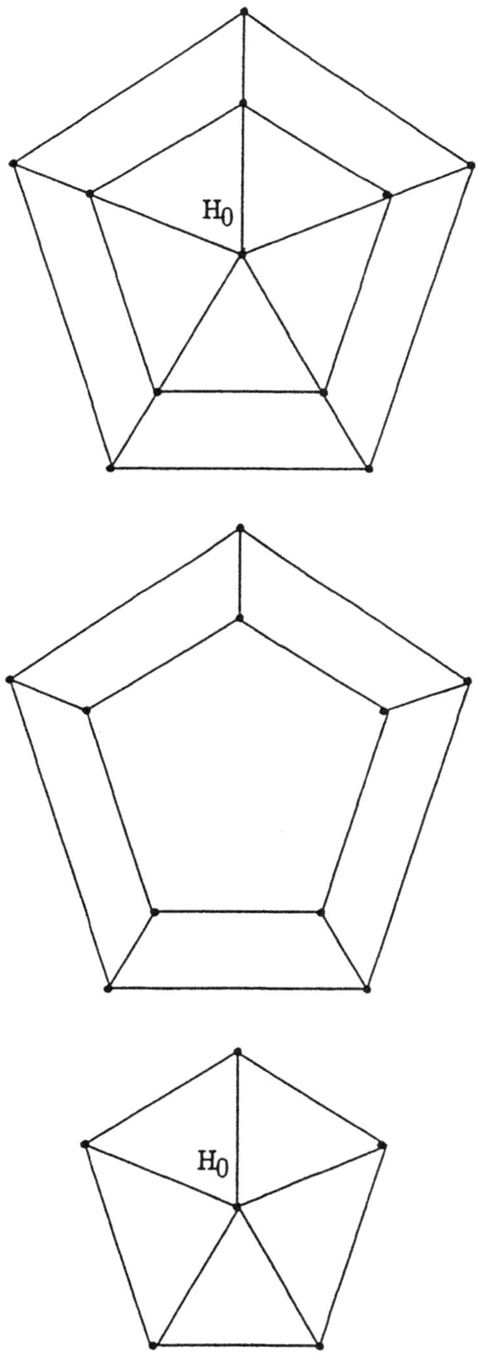

a)

b)

c)

Fig. 3.10. Line graphs for hypergraphs G_k, G_{k-1} and G_k/H_0: a) $L(G_k)$; b) $L(G_{k-1})$; c) $L(G_k/H_0)$.

2. removal and/or contraction of hyperedges of any subset does not make the planar hypergraph nonplanar;
3. if removal (contraction) of hyperedges of some subset $\mathcal{H}_0 \subseteq \mathcal{H}$ of a non-planar hypergraph G makes it planar then removal (contraction) of hyperedges of any subset $\mathcal{H}_1 \supseteq \mathcal{H}_0$ will also make the hypergraph G planar.

Thus we can identify the 'minimal' nonplanar hypergraphs among the set of all nonplanar hypergraphs which are made planar by removal and/or contraction of any its hyperedge. Such hypergraphs are said to be *critical nonplanar.*

Clearly, any nonplanar hypergraph contains a partial hypergraph which is either a critical nonplanar or is made critical nonplanar by contracting hyperedges of a subset.

The set of critical nonplanar hypergraphs of rank 2 (i.e. graphs) is characterized by Theorem 2.4 and consists of two graphs - $K_{3,3}$ and K_5. One can naturally ask what structure has the set of critical nonplanar hypergraphs in a general case. The answer to this question is apparently very difficult to find. This can be supported by the following statement.

Theorem 3.5 *[49] For any $p > 0$ there exists a critical nonplanar hypergraph of rank 3 with the number of vertices and hyperedges being at least p.*

Proof. Such hypergraph G can be constructed as follows.

Consider the graph G_k' obtained from the graph $K_{3,3}$ by splitting its two edges $[v_1, v_6]$ and $[v_2, v_5]$ as shown in Figure 3.11, where $k \geq \frac{p}{3}$.

On the graph G_k' we construct the hypergraph $G_k = (V_k, \mathcal{H}_k)$ by removing the edges $[u_{i2}, u_{i3}]$ and inserting the edges $\{u_{i1}, u_{i3}, u_{i5}\}$, $\{u_{i2}, u_{i3}, u_{i4}\}$, $\{u_{i2}, u_{i3}, u_{i6}\}$ for every $i = 1, ..., k$ (Figure 3.12).

Let us consider three auxiliary statements.

Lemma 3.1 *The hypergraph G_k is not planar.*

Proof of Lemma 3.1. Let us assume that there is some planar realization $R = (V_k, E)$ of the hypergraph G_k. Then for some $1 \leq i \leq k$ the edges $[u_{i2}, u_{i3}] \notin E$, $[u_{i1}, u_{i3}] \notin E$, because otherwise the graph R can be contracted to $K_{3,3}$. Let i_0 be the minimum such value. Then the edges $[u_{i_0 1}, u_{i_0 5}]$, $[u_{i_0 2}, u_{i_0 4}]$, $[u_{i_0 2}, u_{i_0 6}]$, $[u_{i_0 3}, u_{i_0 4}]$, $[u_{i_0 3}, u_{i_0 5}]$, $[u_{i_0 3}, u_{i_0 6}]$, must belong to the graph R, and therefore the subgraph R' in Figure 3.13 can be transformed to the graph $K_{3,3}$ by contracting the chains $(v_1, u_{11}, ..., u_{i_0 1})$, $(v_1, v_4, v_2, ..., u_{i_0 4})$ and $(v_1, v_5, ..., u_{i_0 6})$. This means that the graph R' is not planar. It contradicts the assumption of R planarity. The lemma is proved. □

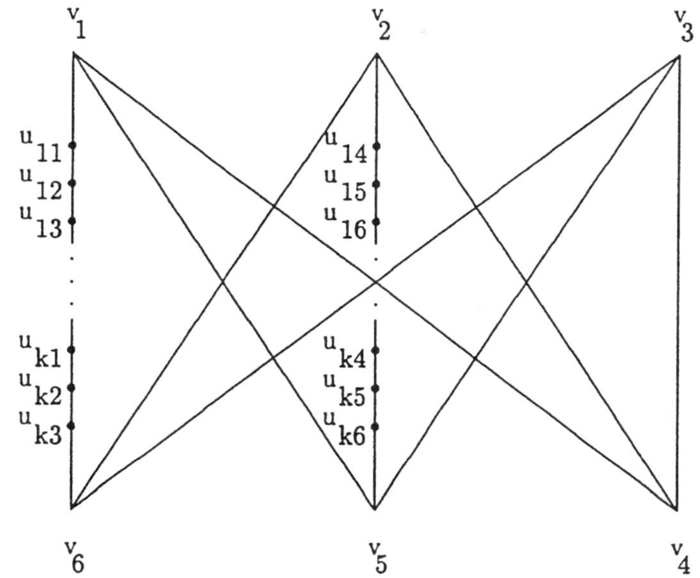

Fig. 3.11. The graph G_k'.

Lemma 3.2 *The hypergraph obtained from G_k by removing any hyperedge of degree 3 is planar.*

Proof of Lemma 3.2. There are three possible cases induced by removing one of the edges $\{u_{i2}, u_{i3}, u_{i6}\}$, $\{u_{i2}, u_{i3}, u_{i4}\}$ and $\{u_{i1}, u_{i3}, u_{i5}\}$ for some $1 \leq i \leq k$. The planar embeddings of some planar realizations for each one of these cases are shown in Figures 3.14- 3.16. The lemma is proved. □

Lemma 3.3 *Hypergraph obtained from G_k by contracting any hyperedge of degree 3 is planar.*

Proof of Lemma 3.3. There are three possible cases induced by contracting one of the edges $\{u_{i2}, u_{i3}, u_{i4}\}$, $\{u_{i2}, u_{i3}, u_{i6}\}$ and $\{u_{i1}, u_{i3}, u_{i5}\}$ for some $1 \leq i \leq k$. The planar embeddings of some planar realizations for each one of these cases are shown in Figures 3.17- 3.19. The lemma is proved. □

Let us continue proving the theorem. Let $G_k'' = (V_k'', \mathcal{H}_k'')$ be a minimal nonplanar partial hypergraph of G_k. According to the Lemmas 3.1 and 3.2 all edges of degree 3 from $\mathcal{H}_k$ belong to $\mathcal{H}_k''$. On the hypergraph G_k'' construct the

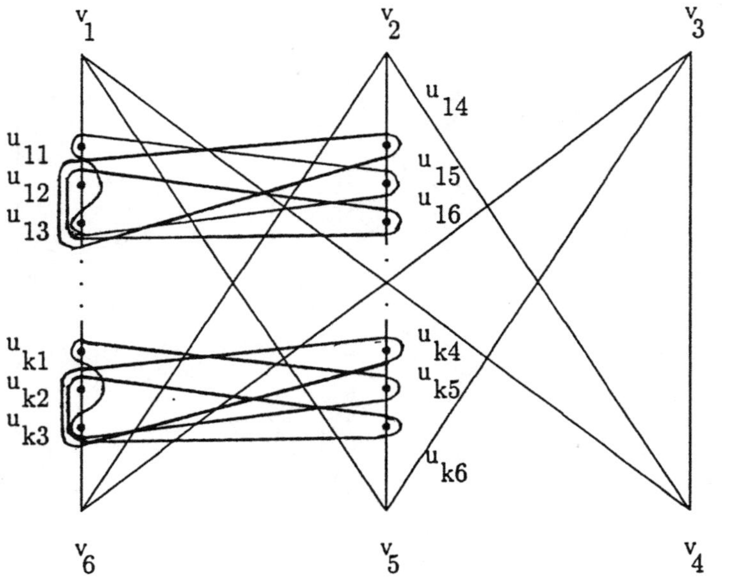

Fig. 3.12. The hypergraph G_k.

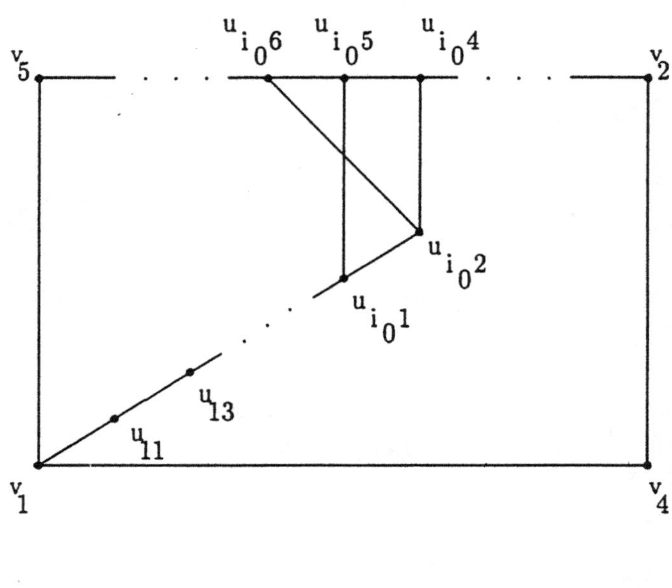

Fig. 3.13. The subgraph R'.

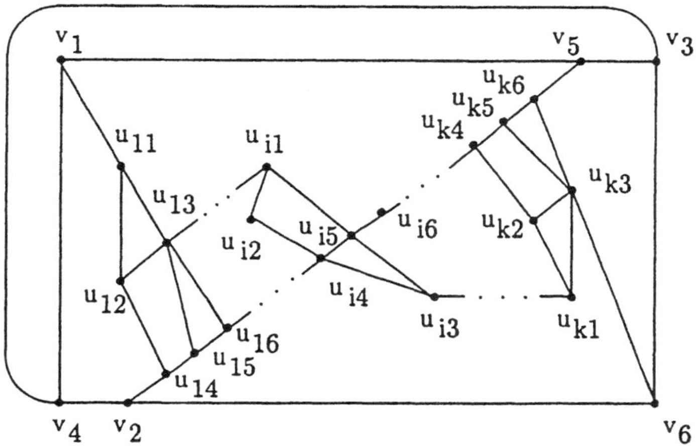

Fig. 3.14. Planar embedding of the realization of the hypergraph obtained from G_k by removing the edge $\{u_{i2}, u_{i3}, u_{i6}\}$.

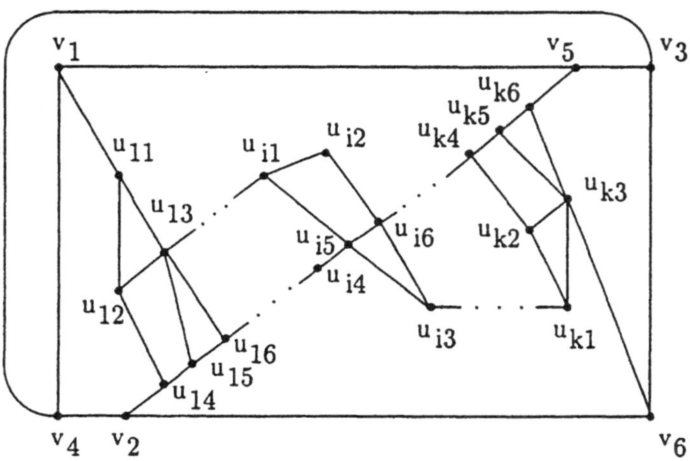

Fig. 3.15. Planar embedding of the realization of the hypergraph obtained from G_k by removing the edge $\{u_{i2}, u_{i3}, u_{i4}\}$.

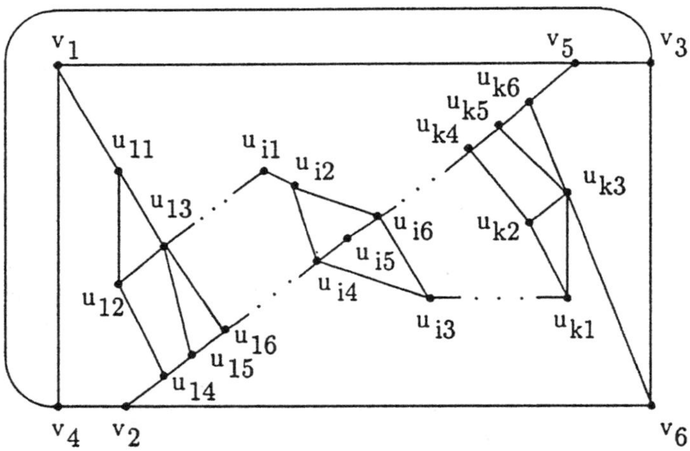

Fig. 3.16. Planar embedding of the realization of the hypergraph obtained from G_k by removing the edge $\{u_{i1}, u_{i3}, u_{i5}\}$.

nonplanar hypergraph $G_k^* = (V_k^*, \mathcal{H}_k^*)$ by contracting the maximum possible number of edges from $\mathcal{H}_k^*$. According to the Lemma 3.3 all edges of degree 3 from $\mathcal{H}_k$ belong to $\mathcal{H}_k^*$. This means that the hypergraph G_k^* meets the theorem's conditions. The theorem is proved. □

In other words, this theorem proves that the set of critical nonplanar hypergraphs (as opposed to graphs) is infinite.

3.2. Planarity of Different Hypergraph Classes

As shown in the previous section the hypergraphs with planar line graphs form one of the planar hypergraph classes. Another one is formed by the so-called *tree-hypergraphs*, i.e. the hypergraphs realized by a tree. The concept of tree-hypergraph is a generalization of that of a tree.

A tree-hypergraph possesses many important properties of a tree. Thus, for example, any tree-hypergraph is 2-colourable. In addition, if an arbitrary hypergraph $G = (V, \mathcal{H})$ is assigned a hypergraph $G^{cl} = (V, \mathcal{H}^{cl})$ where $\mathcal{H}^{cl}$ is a family of all nonempty intersections of hyperedges of $\mathcal{H}$ the following theorem holds.

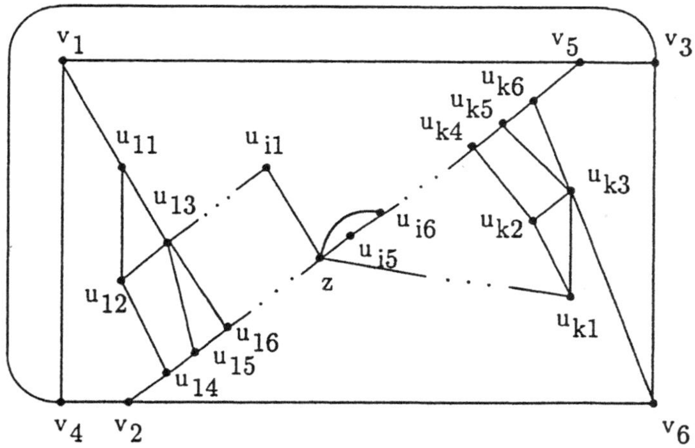

Fig. 3.17. Planar embedding of the realization of the hypergraph obtained from G_k by contracting the edge $\{u_{i2}, u_{i3}, u_{i4}\}$, where $z = \{u_{i2}, u_{i3}, u_{i4}\}$.

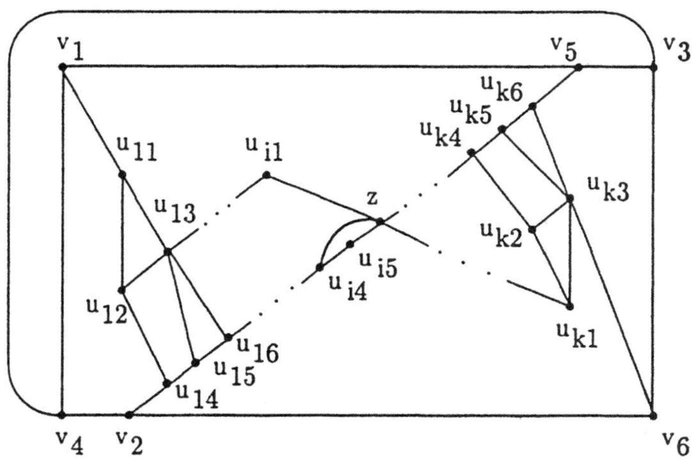

Fig. 3.18. Planar embedding of the realization of the hypergraph obtained from G_k by contracting the edge $\{u_{i2}, u_{i3}, u_{i6}\}$, where $z = \{u_{i2}, u_{i3}, u_{i6}\}$.

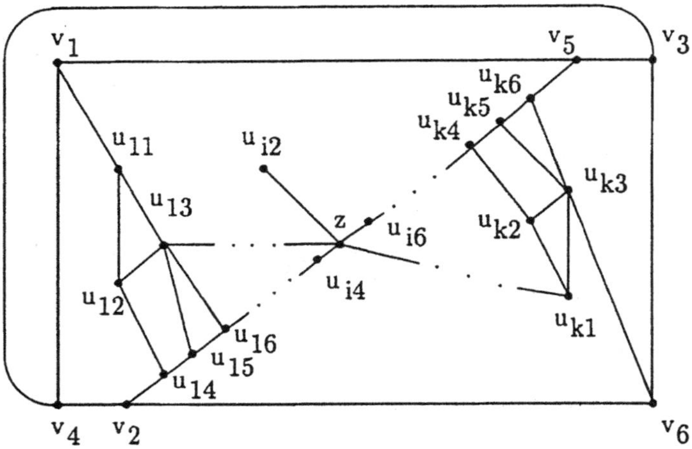

Fig. 3.19. Planar embedding of the realization of the hypergraph obtained from G_k by contracting the edge $\{u_{i1}, u_{i3}, u_{i5}\}$, where $z = \{u_{i1}, u_{i3}, u_{i5}\}$.

Theorem 3.6 *[63] A hypergraph G is a tree-hypergraph if and only if the hypergraph G^{cl} contains no cycles and no three hyperedges of which cross each other.*

Tree-hypergraphs have been well studied. A close relation between tree-hypergraphs and other hypergraph classes has been established.

A hypergraph is said to satisfy the *Helly property* if an intersection of any set of its pairwise intersecting hyperedges is nonempty.

Theorem 3.7 *[29, 74] A hypergraph $G = (V, \mathcal{H})$ is a tree-hypergraph if and only if it satisfies the Helly property and its line graph is chordal.*

Consider a hypergraph $G = (V, \mathcal{H})$. Let $(G)_2$ be a graph whose vertex set coincides with V and that two vertices are adjacent if and only if they are adjacent in the hypergraph G. Any edge $[u, v]$ of the graph $(G)_2$ is assigned the number $\alpha(u, v)$ which is equal to the number of hyperedges from $\mathcal{H}$ containing the edge.

Theorem 3.8 *[63] A hypergraph $G = (V, \mathcal{H})$ is a tree-hypergraph if and only if the set $M(G)$ of minimum (with respect to the number of edges) realizations* *of hypergraph G coincides with the set of all minimum by weight (with respect to α) spanning trees of graph $(G)_2$.*

From this theorem there follows an algorithm testing whether the hypergraph G is a tree-hypergraph.

1. Construct a minimal by weight (with respect to α) spanning tree T of graph $(G)_2$.
2. If a tree T is a realization of a hypergraph G then G is naturally a tree-hypergraph. Otherwise, the hypergraph G is not a tree-hypergraph.

The time complexity of this algorithm does not exceed $O(mn^2)$.

Tree-hypergraphs form a rather narrow class of planar hypergraphs. Therefore it is only natural to relax the constraint of tree realization so that the hypergraph remains planar.

A hypergraph $G = (V, \mathcal{H})$ is said to satisfy the *relaxed Helly property* if for any set $\mathcal{H}_0$ of pairwise intersecting hyperedges from $\mathcal{H}$ there exist two vertices $a, b \in V$ such that $\{a, b\} \cap H \neq \emptyset$ for all $H \in \mathcal{H}_0$.

Let $\mathcal{H}_1, ..., \mathcal{H}_k$ be all maximal sets of pairwise intersecting hyperedges of the hypergraph $G = (V, \mathcal{H})$; let a_1, b_1 be vertices from V such that $\{a_1, b_1\} \cap H \neq \emptyset$ for all $H \in \mathcal{H}_i$, $i = 1, ..., k$. We shall denote by $\tilde{G} = (V, \tilde{\mathcal{H}})$ a hypergraph obtained by adding edges $\{a_i, b_i\}$, $i = 1, ..., k$ to G.

Theorem 3.9 *[51] Let a hypergraph G satisfy the relaxed Helly property and its line graph $L(G)$ be chordal. Then a hypergraph $\tilde{G}$ and hence a hypergraph G are planar.*

To prove this theorem consider the following auxiliary problem. Given a tree $T = (X, A)$ and a nonempty subset $X_0 \subseteq X$. Make a graph $R(T, X_0)$ by replacing every vertex $v \in X_0$ with a pair of adjacent vertices x' and x'', every edge $[x, y] \in A$, $x \in X_0$, $y \in X \backslash X_0$ with pair of edges $[x', y]$ and $[x'', y]$, every edge $[x, y] \in A$, $x, y \in X_0$ with four edges $[x', y']$, $[x'', y']$, $[x', y'']$ and $[x'', y'']$.

Lemma 3.4 *Graph $R(T, X_0)$ is planar.*

Proof. To prove the lemma we will use the mathematical induction on $|X_0|$. The lemma is obviously true for $|X_0| = 0, 1$.

Now let us assume that the graph $R(T, X_0)$ is planar for $|X_0| \leq p - 1$, $p \geq 2$. Let us prove the planarity of the graph $R(T, X_0)$ for $|X_0| = p$. We remove from the tree T the arbitrary vertex $x \in X_0$. As a result the tree T

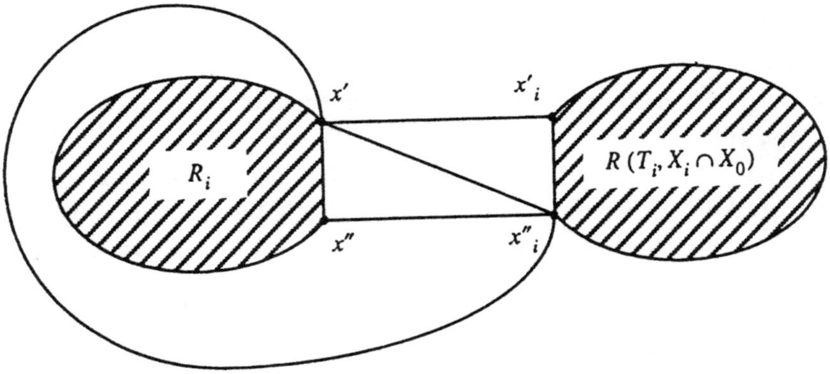

Fig. 3.20. A planar embedding of the graph R_i

will split into a set of subtrees $T_1 = (X_1, A_1),..., T_k = (X_k, A_k)$. Every tree T_i contains a single vertex adjacent to x in the tree T.

According to the induction hypothesis the graphs $R(T_i, X_i \cap X_0)$ are planar for all $1 \leq i \leq k$.

Let $x_1, ..., x_k$ be the vertices adjacent to x in the tree T, where the vertex $x_i \notin X_0$ for $1 \leq i \leq q$ and the vertex $x_i \in X_0$ for $q+1 \leq i \leq k$. We denote by R_0 the graph $(\{x', x"\}, \{[x', x"]\})$. Then we denote by R_i for $i = 1, ..., q$ the graph obtained by adding the edges $[x', x_i]$, $[x", x_i]$ to the graph $R_{i-1} \cup R(T_i, X_i \cap X_0)$, and for $i = q+1, ..., k$ the graph obtained by adding the edges $[x', x'_i]$, $[x', x"_i]$, $[x", x'_i]$, $[x", x"_i]$ to the graph $R_{i-1} \cup R(T_i, X_i \cap X_0)$. The planarity of graphs R_i for $i = 1, ..., q$ is obvious.

Using stereographic projection let us make one face of the graph R_{i-1}, $i = q+1, ..., k$, containing the edge $[x', x"]$ and one face of the graph $R(T_i, X_i \cap X_0)$, containing the edge $[x'_i, x"_i]$ external. You can see the planarity of graph R_i from the Figure 3.20.

Obviously $R(T, X_0) = R_k$. The lemma is proved. □

Proof of the theorem 3.9. Without loss of generality we can assume that the hypergraph G is connected. Let $a_i \neq b_i$ for $1 \leq i \leq l$ and $a_i = b_i$ for $l+1 \leq i \leq k$. If $l = 0$ then according to the Theorem 3.7 the hypergraph G is a tree-hypergraph, and hence it is planar.

Let us take $l > 0$. We construct a new hypergraph $G^* = (V^*, \mathcal{H}^*)$ by inserting for every $1 \leq i \leq l$ a new vertex c_i into every edge from $\mathcal{H}_i$. The hypergraph G^* satisfies the Helly property and its line graph $L(G^*)$ is chordal.

We will prove that the hypergraph G^* can be realized by a tree T^* in which every vertex a_i and b_i is adjacent to the vertex c_i for all $1 \leq i \leq l$. To prove this we will use induction on $|V^*|$ similar to the tree-hypergraph criterion proof of [74]. For $|V^*| = 3, 4$ the statement is obviously true. Let us assume that the statement is true for $|V^*| \leq n - 1$ and prove its correctness for $|V^*| = n$.

Since the line graph $L(G^*)$ is chordal, there is an edge $H_1^* \in \mathcal{H}^*$ such that any edges adjacent to H_1^* are adjacent to each other. Let H_1^* be adjacent only to the edges H_j^*, $j = 2, ..., p$. Since the hypergraph G^* meets the Helly property, there is the vertex $c_i \in \bigcap_{j=1}^{p} H_j^*$.

Let us assume $H_1^* = \{a_i, b_i, c_i\}$. Set $H_j' = H_j^* \backslash \{a_i, b_i\}$ for $j = 2, ..., m$, where $m = |\mathcal{H}|$, $V' = V^* \backslash \{a_i, b_i\}$, $\mathcal{H}' = \{H_j' : j = 2, ..., m\}$, $G' = (V', \mathcal{H}')$. It is obvious that for $H_{j_1}^*, H_{j_2}^* \in \mathcal{H}^*$ the set $H_{j_1}' \cap H_{j_2}' \neq \emptyset$ if and only if $H_{j_1}^* \cap H_{j_2}^* \neq \emptyset$. This means that the line graph $L(G')$ is chordal and the hypergraph G' meets the Helly property. Since $| V' | = | V^* | - 2$ then there is a realization of G' by the tree T' in which every vertex a_i and b_i is adjacent to the vertex c_i for all $1 \leq i \leq l$, $j \neq i$. We can get the required realization T^* by adding to T' two new vertices a_i, b_i and edges $[a_i, c_i]$ and $[b_i, c_i]$.

Now let us assume $H_1^* \neq \{a_i, b_i, c_i\}$. In this case there is a vertex $v \in H_1^* \backslash \{a_i, b_i, c_i\}$. We make a new hypergraph G' by removing the vertex v from hypergraph G^*. It is easy to see that the line graph $L(G')$ is chordal and that the hypergraph G' meets the Helly property. According to the induction hypothesis there is a realization of G' by a tree T' in which every vertex a_i and b_i is adjacent to the vertex c_i for all $1 \leq i \leq l$. We can get the required realization T^* by adding to T' the new vertex v and edge $[v, c_i]$.

Let $C = \{c_i : i = 1, ..., l\}$. We construct the graph R by contracting in the graph $R(T^*, C)$ the edges $[a_i, c_i']$ and $[b_i, c"_i]$ for $i = 1, ..., l$ and removing the edges which are not a subset of any edge from $\tilde{\mathcal{H}}$. The constructed graph R is the realization of the hypergraph $\tilde{G}$. According to Lemma 3.4 the graph $R(T^*, C)$ is planar and, consequently, the graph R is planar too. The theorem is proved. □

We would like to note that the theorem's conditions can not be relaxed. Thus, there is a nonplanar hypergraph with chordal line graph, every hyperedge of which intersects some fixed three- element set of vertices. As an example of such hypergraph we can consider the hypergraph $G = (V, \mathcal{H})$, where $V = \{1, 2, ..., 12\}$, $\mathcal{H} = \{\{1, 2, 3, 10\}, \{4, 5, 6, 10\}, \{7, 8, 9, 10\}, \{1, 4, 7, 11\},$

$\{2,5,8,11\}$, $\{3,6,9,11\}$, $\{1,5,9,12\}$, $\{2,6,7,12\}$, $\{3,4,8,12\}$, $\{1,6,8,10\}$, $\{2,4,9,10\}$, $\{1,6,8,11\}$, $\{2,4,9,11\}$, $\{1,6,8,12\}$, $\{2,4,9,12\}$, $\{10,11,12\}\}$. It easy to see that $L(G) = K_{16}$ and consequently $L(G)$ is chordal. Moreover for any $H \in \mathcal{H}$ the set $H \cap \{10,11,12\} \neq \emptyset$.

Now let us prove that this hypergraph is not planar. Let assume that there is a planar realization $R = (V,E)$ of it. Then there are seven vertices from the set $\{1,2,...,9\}$ every one of which is adjacent to at least one vertex from the set $\{10,11,12\}$ (otherwise the graph R contains $K_{3,3}$ as subgraph). At least two of these seven vertices belong to the set $\{1,6,8,2,4,9\}$. Therefore in the set E there are at least two edges from the set $(\{1,6,8\} \times \{1,6,8\}) \cup (\{2,4,9\} \times \{2,4,9\})$. This means that $|E| \geq 9 \cdot 3+2+2 = 31 > 3 \cdot 12-6$. This contradicts to the planarity of R, since for planar graphs $|E| \leq 3 \cdot |V| - 6$. Consequently, the hypergraph G is not planar.

In the paper [40] D.Johnson and H.Pollack introduced a class of edge-planar hypergraphs. A hypergraph $G = (V, \mathcal{H})$ is called an *edge-planar* if there exists a planar graph $R = (\mathcal{H}, F)$ such that for any subset $V' \subseteq V$ a subgraph of R induced by the vertex set $\{H \in \mathcal{H}; V' \subseteq H\}$ is planar. There is a close relation between planar and edge-planar hypergraphs.

Theorem 3.10 *[40] A dual hypergraph of an edge-planar hypergraph is planar.*

Theorem 3.11 *[40] If a hypergraph G^{cl} is planar then the hypergraph G^* dual to G is edge-planar.*

In the general case the problem of hypergraph planarity testing is NP-complete [3, 40]. However, as we will show below, the solution of some VLSI planarization problems is reduced to the problem of hypergraph planarization without odd-length cycles and with the degrees of all vertices not exceeding 2. It turns out, however, that the hypergraph planarity testing problem is also NP-complete in this case.

Theorem 3.12 *A planarity testing problem for hypergraphs without odd-length cycles and with the degrees of all vertices not exceeding 2 is NP-complete.*

Proof. It is evident that the problem belongs to the NP class. We shall show that a well-known NP-complete problem about the existence of a Hamiltonian path in a 3-connected cubic planar graph [30] can be reduced to that problem.

Let $R = (V,E)$ be a 3-connected cubic planar graph. The proposed reduction will be implemented in two steps. In the first step, as in [40], on the graph R we shall construct a hypergraph G^* whose planarity is equivalent to the existence of a Hamiltonian path in the graph R. In the second step, on the hypergraph G^* we shall construct a hypergraph G which satisfies the conditions of the theorem and which is planar if and only if the hypergraph G^* is planar.

Next we shall take a plane embedding of graph R and assume that $F = \{f_1, ..., f_m\}$ is its face set. Construct a new graph

$$R' = (V \cup F, E \cup E'),$$

where $E' = \{[f, v] : v \text{ is the vertex of the boundary of face } f\}$.

Since the graph R is a 3-connected graph R' is determined uniquely. Replace each edge $e = [u, v] \in E$ with a triple of edges $[u, x_1(e)]$, $[x_1(e), x_2(e)]$, $[x_2(e), v]$ and each edge $e = [u, v] \in E'$ with a pair of edges $[u, x_1(e)], [x_1(e), v]$ where $x_1(e)$ and $x_2(e)$ are new vertices. The resulting graph will be denoted by $R" = (V", E")$.

Then on the graph $R"$ we shall construct a hypergraph $G^* = (V^*, \mathcal{H}^*)$ assuming that

$$V^* = V^" \cup V^',$$

where V' is a set containing a new vertex v for each vertex $v \in V$ and

$$\mathcal{H}^* = E' \cup \{\{V \cup V'\}\} \cup \{G_v : v \in X\},$$

where $C_v = \{v'\} \cup N_v$ and N_v is the vertex set from $V"$ which are adjacent to v in graph $R"$.

As has been shown in [40], the planarity of the hypergraph G^* is equivalent to the existence of a Hamiltonian path in graph R.

Now we shall proceed to the second step of reduction.

Consider a set V_0 of vertices from $V^"$ whose degree is greater than 2 in the hypergraph G^*. Let v be an arbitrary vertex from V_0. We shall denote by $\mathcal{N}(v)$ the set of hyperedges from $\mathcal{H}^*$ which are incident on the vertex v. Construct a hypergraph $G = (X, \mathcal{H})$ letting

$$X = (V^* \backslash (V \cup F)) \cup \{u(v, H) : v \in V_0, H \in \mathcal{N}(v)\},$$

$$\mathcal{H} = (\mathcal{H} \backslash (\bigcup_{v \in V_0} \mathcal{N}(v))) \cup \mathcal{H}' \cup \mathcal{H}^",$$

where

$$\mathcal{H}' = \{(H \backslash \{v\}) \cup \{u(v, H)\} : v \in V_0, H \in \mathcal{N}(v)\},$$

$$\mathcal{H}'' = \{\{u(v,H) : H \in \mathcal{N}(v)\} : v \in V_0\}.$$

It is easy to see that the degree of each vertex in the hypergraph G is 2 and the hypergraph G contains no odd-length cycles and is planar if and only if the hypergraph G^* is planar. □

The method of proof uses significantly the infinity of hypergraph rank. Hence the question of whether the problem is NP-complete remains open when the hypergraph rank is restricted to a constant. Meanwhile such problems are often encountered in practice.

Theorem 3.13 *[4] The problem of planarity testing for hypergraphs whose rank does not exceed 8 is NP-complete.*

When the hypergraph rank is 3, 4, 5, 6 or 7 the question about NP-completeness remains open.

Consider some cases where the problem of hypergraph planarity is solved with polynomial time complexity.

Let $G_i = (V_i, E_i)$, $i = 1, \ldots, k$ be vertex-disjoint 3-connected planar graphs. Construct a hypergraph G by adding a hyperedge set $\mathcal{H}'$ for each hyperedge H of which $| V_i \cap H | \leq 1$, $i = 1, \ldots, k$ to the graph $\bigcup_{i=1}^{k} G_i$.

On the hypergraph G, we shall construct a graph $\Gamma(G)$ by adding a vertex $v(H)$ and edges $[v(H), u]$ to the graph $\bigcup_{i=1}^{k} G_i$ where $u \in H$ for each $H \in \mathcal{H}'$. Clearly, a hypergraph G is planar if and only if the graph $\Gamma(G)$ is planar.

For a hypergraph $G = (V, \mathcal{H})$ we shall denote the set of hyperedges of degree 2 from $\mathcal{H}$ by $\mathcal{H}^{(2)}$ and the graph $(V, \mathcal{H}^{(2)})$ by $G^{(2)}$. Suppose the rank of the hypergraph G is 3 and the graph $G^{(2)}$ is planar and 3-connected. In that case the problem the planarity of the hypergraph G is solved in polynomial time [2]. Let $\mathcal{H}^{(1)}$ be the set of hyperedges of degree 3 from $\mathcal{H}$. Without loss of generality we may assume that $\mathcal{H}' = \{H_i = \{x_i, y_i, z_i\} : i = 1, ...t\}$. The graph $R = (V, \mathcal{H}^{(2)} \cup E)$ is a realization of the hypergraph G if it containts at least two edges from each triple $T_i = \{[x_i, y_i], [y_i, z_i], [z_i, x_i]\}$. We shall assign a Boolean variable α_j which becomes 1 if $e_j \in E$ and 0 otherwise to each edge e_j from the set $T = \bigcup_{i=1}^{t} T_i = \{e_j : j \in J\}$. Each triple $T_i = \{e_j, e_k, e_l\}$ is assigned a collection of disjunctions $S_i = \{\alpha_j \vee \alpha_k, \alpha_k \vee \alpha_l, \alpha_l \vee \alpha_j\}$. Thus the graph R is a realization of the hypergraph G if and only if E consists of those e_j which correspond to a satisfying truth assignment for the set of disjunctions $S = \bigcup_{i=1}^{t} S_i$.

Since the graph $G^{(2)}$ is 3-connected the addition of an edge from E does not affect planarity only if the vertices connected by this edge belong to the same face.

To account for the planarity condition, we must extract a set $K = \{[e_{ip}, e_{jp}] : p = 1, ..., k\}$ of all conflicting edge pairs from T with regard to the faces of graph $G^{(2)}$. The planarity of R is equivalent to the existence of at most one edge from each pair of K in E. The latter corresponds to satisfiability of the set of disjunction $B = \bigcup_{i=1}^{k} \{\alpha'_{ip} \vee \alpha'_{jp}\}$.

Thus the hypergraph planarity testing problem is reduced to the problem of finding the satisfaction of the truth assignment on the set $\{\alpha_j\}$ for the disjunction set $S \cup B$. Note that each disjunction from $S \cup B$ contains not more than two literals. We have obtained a well-known 2-satisfiability problem to solve which there is a polynomial algorithm [30]. Its running time does not exceed $O(|I||S \cup B|)$ in our case. The time complexity of constructing the set K does not exceed $O(fm^2)$, where f is the number of faces in the graph $G^{(2)}$ and m is the number of hyperedges in the hypergraph G. Hence the time complexity of planarity testing for the hypergraph G does not exceed $O(nm^2 + m^3)$, where n is the number of vertices in the hypergraph.

Now consider a hypergraph $G = (V, \mathcal{H})$ which contains no more than one vertex in the intersection of any two hyperedges. Let the graph $G^{(2)}$ be planar and 3-connected as earlier, $E(G^{(2)})$ be the set of edges in graph $G^{(2)}$ whose ends belong to the same face of $G^{(2)}$, and K be the collection of edge pairs from $E(G^{(2)})$ which conflict with regard to the faces of graph $G^{(2)}$. Assume that no edge in $E(G^{(2)})$ enters more than one pair from K.

Let us replace each pair of edges $(e_i = [x_i, y_i], e_j = [x_j, y_j]) \in K$ with five edges $e_i^1 = [x_i, a_{ij}]$, $e_i^2 = [y_i, b_{ij}]$, $e_j^1 = [x_j, a_{ij}]$, $e_j^2 = [y_j, b_{ij}]$ and $e_{ij} = [a_{ij}, b_{ij}]$ (Figure 3.21). We denote the resulting graph by $G = (X, E)$. Each hyperedge $H_i \in \mathcal{H} \backslash \mathcal{H}^{(2)}$ is assigned a subgraph $P_i = (X_i, E_i)$ of graph G induced by the vertex set

$$X_i = \bigcup_{(p,q) \in I_i} \{a_{pq}, b_{pq}\} \cup H$$

where I_i is the set of index pairs for such conflicting edges one of which is contained in the hyperedge $H_i \in \mathcal{H}$.

It may be shown [2] that a hypergraph G is planar if and only if there exists a set of trees $\{T_i\}$ which have no common edges each of which is a spanning tree for the appropriate graph P_i. Let $\{T_i\}$ be such set of trees. We shall construct a set of trees $\{T_i'\}$ by letting $E(T_i') = (E(G^{(2)}) \cap E(T_i)) \cup \{e_p = [x_p, y_p] : x_p, y_p \in H_i$ and q such that $[a_{pq}, b_{pq}] \in E(T_i)\}$ where $E(T_i')$ and $E(T_i)$ are the sets of edges for tree T_i' and T_i, respectively. It can be readily verified that the time complexity of reducing the hypergraph planarity problem to the problem of finding the set of edge-intersecting trees does not

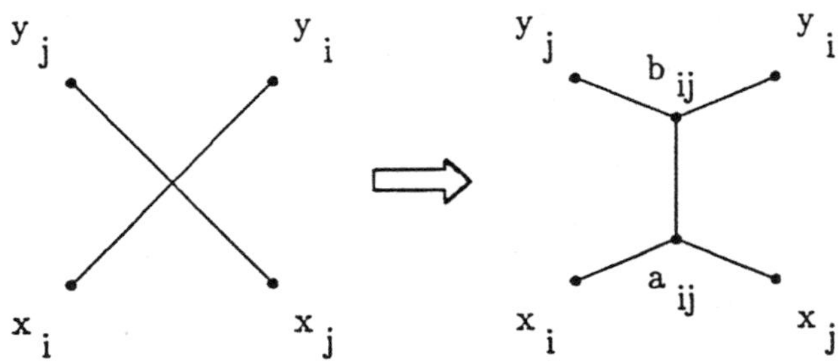

Fig. 3.21.

exceed $O(|E(G^{(2)})||K|)$ in this case. A polynomial algorithm for the latter problem can be found in [69].

3.3. Hypergraph Nonplanarity Measures. Statement of Hypergraph Planarization Problem

As noted in item 2.1, the number of edges in a planar graph does not exceed $3n - 6$ where n is the number of vertices. At the same time, the number of edges in the majority of graphs is greater by far (it proportial to n^2). This suggests, as was discussed in [43], that almost all graphs are nonplanar, i.e. if we denote the set of all graphs on n vertices by $\Im_2(n)$ and the set of all planar graphs on n vertices by $\Im_2^p(n)$ then

$$\lim_{n\to\infty} \frac{|\Im_2^p(n)|}{|\Im_2(n)|} = 0.$$

It turns out that a similar result exists for hypergraphs. We denote the set of all hypergraphs on n vertices by $\Im(n)$ and the set of all planar hypergraphs from $\Im(n)$ by $\Im^p(n)$. Then the following holds.

Proposition 3.3 *Almost all hypergraphs are nonplanar, i.e.,*

$$\lim_{n\to\infty} \frac{|\Im^p(n)|}{|\Im(n)|} = 0.$$

Proof. Any hypergraph $G = (V, \mathcal{H}) \in \Im(n)$ can be presented uniquely as the union of a graph G_1 and a hypergraph G_2 for which degrees of all hyperedges exceed 2. Thus the set $\Im(n)$ can be represented as a Cartesian product of set $\Im_2(n)$ and set $\Im'(n)$ of hypergraphs from $\Im(n)$ for which the degrees of all hyperedges exceed 2. Hence $|\Im(n)|=|\Im_2(n)||\Im'(n)|$ and

$$\frac{|\Im^p(n)|}{|\Im(n)|} < \frac{|\Im'(n)||\Im_2^p(n)|}{|\Im(n)|} = \frac{|\Im_2^p(n)|}{|\Im_2(n)|}.$$

Since

$$\lim_{n\to\infty} \frac{|\Im_2^p(n)|}{|\Im_2(n)|} = 0.$$

then

$$\lim_{n\to\infty} \frac{|\Im^p(n)|}{|\Im(n)|} = 0.$$

The proposition is proved. □

Therefore the majority of graphs and hypergraphs obtained in practice is nonplanar. In this connection, it would be of interest to estimate the degree of nonplanarity, i.e. to what extent the graph or hypergraph in question differs from a planar one.

Consider five topological invariants of a graph which characterize different nonplanarity properties.

Genus $\gamma(G)$ of a graph G is the smallest number of handles to be added to a sphere such that the graph G be embeddable on the obtained surface. We can readily show that if $G_1, ..., G_k$ are the blocks of graph G then

$$\gamma(G) = \sum_{i=1}^{k} \gamma(G_i).$$

We present some well-known estimates of graph genus. Clearly, if a graph G has $n > 3$ vertices and m edges then

$$\gamma(G) \geq \left[\frac{1}{6}(m - 3n) + 1\right].$$

The genus of the complete graph is

$$\gamma(K_n) = \left]\frac{(n-3)(n-4)}{12}\right[, n \geq 3.$$

The genus of the complete bipartite graph is

$$\gamma(K_{p,q}) = \left]\frac{1}{4}(p-2)(q-2)\right[.$$

The genus of the n-dimensional cube is

$$\gamma(Q_n) = 1 + (n-4)2^{n-8}.$$

The *thickness* $t(G)$ of a graph G is the minimum number of its planar subgraphs the union of which produces the graph G. For the thickness of a connected graph with n vertices and m edges the following estimates hold true:

$$t(G) \geq \left]\frac{m}{3n-6}\right[, t(G) \geq \left[\frac{m+3n-7}{3n-6}\right].$$

From this the following thickness estimate is obtained for a complete graph :

$$t(K_n) \geq \left[\frac{n+7}{6}\right].$$

It turns out that if $n \neq 4(\mathit{mod}\ 6)$ and $n \neq 9$ then

$$t(K_n) = \left[\frac{n+7}{6}\right].$$

For a complete bipartite graph,

$$t(K_{m,n}) = \left\{\frac{mn}{2(m+n-2)}\right\},$$

except possibly the cases when $m < n, mn$ is odd and there exists an integer k such that

$$n = \left[\frac{2k(m-2)}{m-2k}\right].$$

Hence

$$t(K_{n,n}) = \left\{\frac{n+5}{4}\right\}.$$

For an n-dimensional cube,

$$t(Q_n = \left\{\frac{n+1}{4}\right\}.$$

The *crossing number* $cr(G)$ of a graph G is the smallest number of intersections obtained in the image of the graph on a plane (the notion of intersection refers to intersection of only two edges). The following estimates for the crossing number are known for a complete graph and a complete bipartite graph.

$$cr(K_n) \leq \frac{1}{4}\left[\frac{n}{2}\right]\left[\frac{n-1}{2}\right]\left[\frac{n-2}{2}\right]\left[\frac{n-3}{2}\right],$$

$$cr(K_{p,q}) \leq \left[\frac{p}{2}\right]\left[\frac{p-1}{2}\right]\left[\frac{q}{2}\right]\left[\frac{q-1}{2}\right],$$

The *skewness* $sk(G)$ of a graph G is the smallest number of edges whose removal leads to a planar graph.

To determine the fifth invariant, we need an operation of vertex splitting, i.e. the vertex v is replaced with two new nonadjacent vertices v_1 and v_2 and any edge $[u, v]$ is replaced with the edge $[u, v_1]$ or $[u, v_2]$. The *splitting number* $sp(G)$ of graph G is the smallest number of splits required to transform a nonplanar graph G into a planar graph.

The presented invariants can be readily extended to hypergraphs. The value of a hypergraph invariant is a minimum value among the appropriate realization invariants for the given hypergraph. In other words, if $Rel(G)$ is the set of realizations of the hypergraph G then the *genus* of the hypergraph G is the value

$$\gamma(G) = \min_{R \in Rel(G)} \gamma(R);$$

the *thickness* of the hypergraph G is

$$t(G) = \min_{R \in Rel(G)} t(R);$$

the *crossing number* of the hypergraph G is

$$cr(G) = \min_{R \in Rel(G)} cr(R);$$

the *skewness* of the hypergraph G is

$$sk(G) = \min_{R \in Rel(G)} sk(R);$$

Precise calculation of the above invariants involves great difficulties, even for graphs.

For computer-aided design of VLSI, extraction of the maximal planar part of the circuit is of major importance. The latter problem is reduced to extraction of the maximum planar part of the corresponding hypergraph. A *planar part* of the hypergraph $G = (V, \mathcal{H})$ is a planar hypergraph $G' = (V, \mathcal{H}')$ every hyperedge of which is contained in a hyperedge from $\mathcal{H}$. For the planar part G' we define

$$ins(G, G') = \sum_{H \in \mathcal{H}} (k(H) - 1)$$

where $k(H)$ is the number of connected components in a graph obtained from $(V, \emptyset)$ by adding all edges $[v, v']$ where v and v' belong to a hyperedge $H' \in \mathcal{H}', H' \subseteq H$. Clearly,

$$ins(G, G') \geq sk(G).$$

The planar part G' of hypergraph G with the smallest value of $ins(G, G')$ we shall call the *maximum planar part.*

3.4. Decomposition Methods of Hypergraph Planarization

From a practical standpoint, decomposition algorithms which allow us to reduce the original problem to a collection of similar problems for 'nondecomposable' hypergraphs and then planarize them by the approximation algorithms are the most promising for hypergraph planarization. The approaches discussed in this item allow us to deduce whether an initial hypergraph is planar or not from the planarity or nonplanarity of some complementary hypergraphs of smaller size, and if the initial hypergraph is planar to construct a planar realization of the hypergraph based on the planar realization of 'nondecomposable' hypergraphs by a specific procedure A. Of prime importance is the fact that the same approaches can be used in particular for the approximate construction of realizations with a minimal nonplanarity measure. For this purpose, realizations of complementary hypergraphs with a minimal nonplanarity measure must be 'applied' to the input of the above procedure A.

The first result in planar graph theory is Whitney's theorem [89] which allows us to reduce planarity testing for any graph to planarity testing for a collection of biconnected graphs. The following theorem shows that a similar result can be obtained for hypergraphs.

Theorem 3.14 *Let $G_1, ..., G_k$ be blocks of a hypergraph G. Then*

$$\gamma(G) = \sum_{i=1}^{k} \gamma(G_i),$$

$$t(G) = \max\{t(G_i) : i = 1, ..., k\},$$

$$cr(G) = \sum_{i=1}^{k} cr(G_i),$$

$$sk(G) = \sum_{i=1}^{k} sk(G_i).$$

Corollary 3.2 *A hypergraph is planar if and only if every block of the hypergraph is planar.*

Let $G = (V, \mathcal{H})$ be a biconnected hypergraph and $\{u, v\}$ be its articulation set's vertices. Denote by $G_i'(u,v) = (V_i', \mathcal{H}_i')$, $i = 1, ..., k$, the i-th connected component of the hypergraph obtained by removing vertices u and v from G. Construct a hypergraph $G_i(u,v)$ by adding a hyperedge $\{u,v\}$ to the hypergraph $(V_i, \mathcal{H}_i)$ where $V_i = V_i' \cup \{u,v\}$, $\mathcal{H}_i = \{H \in \mathcal{H}; H \subseteq V_i\}$.

Theorem 3.15 *Let $\{u,v\}$ be an articulation set of a biconnected hypergraph G. The hypergraph G is planar if and only if the hypergraph $G_i(u,v)$ is planar for any $i = 1, ..., k$.*

Proof. *Necessity.* Let $R = (V, E)$ be a planar realization of the hypergraph G. Contract all edges of the graph R which do not lie in the hyperedges from $\mathcal{H}_i$. The resulting graph R_i is a realization of the hypergraph $G_i(u,v)$. The planarity of the graph R implies that the graph R_i is planar for any $i = 1, ..., k$.

Sufficiency. Let R_i, $i = 1, ..., k$ be a planar realization of a hypergraph $G_i(u,v)$. Consider a graph R which is the union of graphs R_i, $i = 1, ..., k$. Clearly, the graph R is planar. Moreover, if the vertices u and v are adjacent in the hypergraph G then R is a realization of the hypergraph G. Otherwise, remove an edge $[u,v]$ from R. The resulting graph is also planar and is a realization of the initial hypergraph G. □

This then implies that it is sufficient for planarity testing to consider the 3-connected hypergraphs.

We now turn our attention to the discussion of hypergraphs whose edge connectivity does not exceed 2. The problem, as applied to graphs, is meaningless, since the vertex connectivity of a graph never exceeds its edge connectivity.

Let $\mathcal{R}$ be a hyperedge articulation set of the hypergraph $G = (V, \mathcal{H})$ and V_i, $i = 1, ..., k$ be a vertex set for the i-th connected component of the hypergraph obtained by removing the hyperedges of the set $\mathcal{R}$ from G. Denote by $G_i(\mathcal{R})$ a subhypergraph of the hypergraph G generated by the vertex set V_i.

Lemma 3.5 *Let $\mathcal{R}$ be a hyperedge articulation set of a hypergraph G and no two hyperedges from $\mathcal{R}$ be adjacent. Then if the hypergraph G is planar the hypergraphs $G_i(\mathcal{R})$, $i = 1, ..., k$ will also be planar.*

Proof. Let $\mathcal{R}$ be a planar realization of the hypergraph G. Remove all edges which belong neither to hyperedges from $\mathcal{R}$ nor to hyperedges of the hypergraph $G_i(\mathcal{R})$ from the graph $\mathcal{R}$. Construct a graph R_i by contracting all edges which do not belong to the hyperedges of hypergraph $G_i(\mathcal{R})$ in the resulting graph. The planarity of the graph R and the properties of planar graphs imply the planarity of R_i. □

Theorem 3.16 *Let H be a bridge of a hypergraph G. Then the hypergraph G is planar if and only if the hypergraphs $G_i(H)$, $i = 1, ..., k$ are planar.*

Proof. The necessity follows from Lemma 3.5. We shall show the sufficiency. Let $R_i, i = 1, ..., k$ be a planar realization of hypergraph $G_i(H)$. In every graph R_i we select a vertex $u_i \in H$. Construct an arbitrary tree T on the vertex set $u_1, ..., u_k$. Clearly, the graph $R = T \cup (\bigcup_{i=1}^{k} R_i)$ is planar. Moreover, the subgraph $R(H_j)$ is connected for any $H_j \in \mathcal{H}$. □

On a hypergraph $G_i(\mathcal{R})$, we construct a hypergraph $G_i^0(\mathcal{R})$ by adding a new vertex v_i and hyperedges $(H \cap V_i) \cup \{v_i\}, H \in \mathcal{R}$.

Lemma 3.6 *Let $\mathcal{R}$ be a cutset of a hypergraph G and no two of its hyperedges be adjacent. Then if the hypergraph G is planar, the hypergraphs $G_i^0(\mathcal{R})$, $i = 1, ..., k$ are also planar.*

Proof. According to the Lemma 3.5, the hypergraph G is planar. Since $\mathcal{R}$ is a cutset there exists a face of a plane embedding of the $G_i(\mathcal{R})$ realization which contains at least one vertex from each set $H \cap V_i, H \in \mathcal{R}$. This implies that the hypergraph $G_i^0(\mathcal{R})$ is planar. □

Theorem 3.17 *Let $\mathcal{R} = \{H_1, H_2\}$ be a cutset of a hypergraph G and $H_1 \cap H_2 = \emptyset$. Then the hypergraph G will be planar if and only if the hypergraphs $G_i^0(\mathcal{R})$, $i = 1, ..., k$ are planar.*

Proof. The necessity follows from Lemma 3.6. We shall show the sufficiency. Let R_i^0 be a planar realization of the hypergraph $G_i^0(\mathcal{R})$. Without loss of generality, we can consider that the degree of the vertices v_i in the graph R_i^0 is 2. Denote by v_{i1} and v_{i2} the vertices which are adjacent to v_i in R_i^0 with $v_{ij} \in H_j$, $j = 1, 2$. Construct a graph R_i by removing the vertex v_i from R_i^0. Consider the graph R obtained by adding the edges $[v_{ij}, v_{(i+1)j}]$ for

$i = 1, ..., k-1$ and $j = 1, 2$ to $\bigcup_{i=1}^{k} R_i$. Clearly, the graph R is planar and its subgraph induced by the vertex set of an arbitrary hyperedge is connected. $\square$

It should be noted that fairly efficient decomposition approaches to hypergraph planarization have not yet been found, unfortunately. Decomposition problems call for further study and development.

The statements discussed above are the necessary and sufficient conditions of planarity when the initial hypergraph planarity is equivalent to that of a collection of complementary hypergraphs of smaller size. At the same time, to find an approximate solution of the planarization problem we can use only sufficient conditions which allow us to draw a conclusion about the planarity of the initial hypergraph from the planarity of some complementary hypergraphs. Consider two conditions of the kind.

Let $G = (V, \mathcal{H})$ be a finite hypergraph with the vertex set V and a family of hyperedges $\mathcal{H}$. Consider a set $\mathcal{M}$ of all minimal (with respect to the partial order $\subseteq$) non-empty intersections of hyperedges from $\mathcal{H}$. Select a vertex x_M in every set $M \in \mathcal{M}$. Consider a hypergraph $G' = (V', \mathcal{H}')$ obtained by removing all vertices other than x_M, $M \in \mathcal{M}$, from the hypergraph G. The next theorem is a generalization of the Theorem 3.1.

Theorem 3.18 *For a hypergraph G, the inequalities*

$$\gamma(G) \leq \gamma(G'),$$

$$t(G) \leq t(G'),$$

$$cr(G) \leq cr(G'),$$

$$sk(G) \leq sk(G')$$

hold.

Proof. Let R' be a realization of the hypergraph G', v be an arbitrary vertex from $V \backslash V'$ and $H_{i_1}, ..., H_{i_p}$ be all hyperedges incident on v from $\mathcal{H}$.

Select a vertex $x_M \in V$ incident on the hyperedges $H_{i_1}, ..., H_{i_p}$ and add the vertex v and edge $[v, x_M]$ to the graph R.

Acting similarly on every vertex from $V \backslash V'$ we will obtain the graph R which is a realization of the hypergraph G. Clearly, $\gamma(R) \leq \gamma(R')$, $t(R) \leq t(R')$, $cr(R) \leq cr(R')$, $sk(R) \leq sk(R')$. $\square$

Let S be an vertex articulation set of an initial hypergraph G, $G_i' = (V_i', \mathcal{H}_i')$, $i \in I(S) = \{1, ..., k\}$ be the connected components of a hypergraph obtained from G by removing the vertex set S, L_i be the line graph of hypergraph G_i'; $\mathcal{H}_i = \{H \in \mathcal{H} : H \subseteq V_i \cup S\}$; $V_i = V_i' \cup S$; $G_i = (V_i, \mathcal{H}_i)$. Since the hypergraph G_i is a partial hypergraph of G for any $i = 1, ..., k$, the planarity of the hypergraph G implies that the hypergraphs G_i are planar. The reverse is generally false. As an example consider a graph $K_{3,3}$ whose articulation set consists of the vertices of one part.

For each $i = 1, ..., k$ we shall construct a hypergraph $D_i(S)$ by adding to a hypergraph

$$G_i \cup \Big(\bigcup_{\substack{j=1,\dots,k \\ j \neq i}} L_j \Big),$$

for each $j = I(S) \backslash \{i\}$, hyperedges $(H \cap S) \cup \{h\}$ for all $H \in \mathcal{H}$, such that $H \cap S \neq \emptyset$, $H \backslash S \neq \emptyset$ where h is the vertex of graph $\bigcup_p L_p$ associated with the edge $H \backslash S \in \mathcal{H}_j'$.

Theorem 3.19 *[50] For a hypergraph G to be planar, it is sufficient that there is a vertex articulation set S such that the hypergraph $D_i(S)$ is planar for some $i \in I(S)$.*

Chapter 4

Mathematical Models for the VLSI Planarization Problem

The temptation of creating a general-purpose graph model for a circuit adequate to the planarization problem has already long been experienced by the designers of CAD systems. This would allow us to apply the efficient graph algorithms of Chapter 2 for at least circuit planarity testing. It is interesting to note that investigations in this direction had been started well before the highly efficient Hopcroft-Tarjan, Rubin algorithm and other were reported. However, numerous investigations revealed that such a general purpose graph model does not exist in practice for an arbitrary electrical circuit. Moreover it has been ascertained that to planarize a wide class of circuits one must turn to hypergraph models.

4.1. Graph versus Hypergraph VLSI Models

There are two basic approaches to constructing graph models of a circuit. In the first approach vertices represent circuit elements, in the second they represent pins. In both cases a plane implementation of the circuit is naturally associated with a plane graph whose edges correspond to routes. Thus each plane circuit is associated with some plane graph. The prime object of most investigations has been to construct a graph whose planarity is equivalent to that of the circuit for each circuit. As will be revealed later, it is impossible. The major reason is the indefinite representation of a net. Generally, each net is represented by a tree. Tree selection has a pronounced effect on the possibility of planar circuit realization. It is well known that the number of distinct trees on the same vertex set is as large as their scanning becomes prohibitive (n^{n-2} where n is the number of vertices). One can not also be restricted to some specific tree configuration, since this might lead to a false conclusion about circuit nonplanarity. Nevertheless graph models for the circuit planarization problem have gained wide acceptance. This might be attributed primarily to the existence of fairly simple and efficient algorithms for their processing.

A most simple and efficient graph model for the circuit planarization problem is the so called model $\mathbf{M}_1$ [61] which is a bipartite graph. Vertices of the first part correspond to circuit elements and vertices of the second part correspond to its nets. Two vertices are adjacent if some pin of the element associated with one vertex belongs to a net associated with another vertex. The model $\mathbf{M}_1$ is adequate for a circuit whose elements have no physically equivalent pins but all pins are logically equivalent.

Planarity of $\mathbf{M}_1$ is generally neither a necessary nor a sufficient condition for circuit planarity. The first follows from the fact that $\mathbf{M}_1$ implements star-configured nets, the latter from the requirement for logical equivalence of all pins of the element.

The hypergraph model $\mathbf{C}_1$ of a circuit takes account of arbitrary configuration of nets. The model vertices correspond to circuit elements and its hyperedges to its nets. The graph model $\mathbf{M}_1$, by definition, is the Konig representation of the hypergraph model $\mathbf{C}_1$, therefore $\mathbf{M}_1$ planarity implies that $\mathbf{C}_1$ is also planar. The reverse is evidently false.

Let us consider a circuit embedded in the plane as a plane graph R. In this graph circuit elements correspond to the vertices and interconnections of elements correspond to the edges. Then each net in the circuit is associated with a subtree of R where the vertex set coincides with the set of circuit elements involved in the net. Clearly, the graph R is a planar realization of the hypergraph which is model $\mathbf{C}_1$. Thus if the circuit is planar then the model $\mathbf{C}_1$ is evidently planar, too. The reverse is generally false. This may be attributed in particular to the fact that the resultant order of routing connections to an element does not correspond to the arrangement of pins on this element boundary. Planarization of $\mathbf{C}_1$ as an arbitrary hypergraph assumes that each logical pin can be represented by a group of physically equivalent pins that are arbitrarily arranged on the element boundary.

Thus the model $\mathbf{C}_1$ does not account for peculiarities of the elements and is rather just the model of connections in the circuit itself. The circuit model is the union of the connection model and the element model which account for the order of pins, the logical and physical equivalence of pins, as well as other properties of the elements. Different element models are discussed in

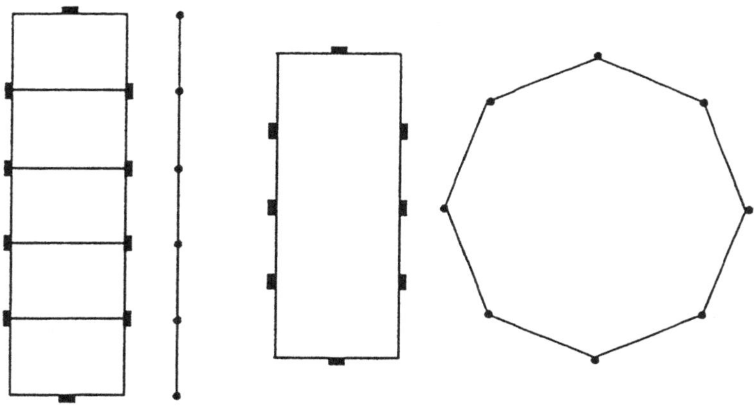

a b

Fig. 4.1. Linear (a) and cyclic (b) elements and their representation

detail in the next item.

Futher on we shall use the model $\mathbf{C}_1$ as a connection model.

4.2. VLSI Models with Specified Element Models

In top-down layout design, planarization of the *i*-th level cell is carried out with the already designed layouts of the *(i+1)*-th level cells. These cells (elements) are associated with the models which allow us to account for the arrangement of logically equivalent groups of pins, logically and physically equivalent pins, metric requirements for over-the-element routes, etc., to a greater or lesser extent.

One such model is the graph model of an element. Here we may distinguish between linear and cyclic elements. Examples of linear and cyclic elements are given in Figure 4.1.

It is good practice to represent a cyclic element with the specified arrangement of pins on its boundary as a cycle whose vertices correspond to physical pins of the element, and edges provide for the required order of their placement. In this case it is necessary to prevent other elements from embedding in the internal face of the cycle. This is obtained by introducing the so called central vertex which is connected to all vertices in the cycle (Figure 4.2).

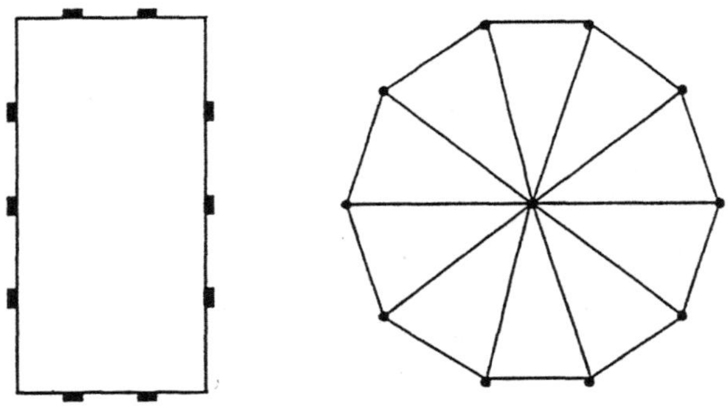

Fig. 4.2. A cyclic element and its graph model

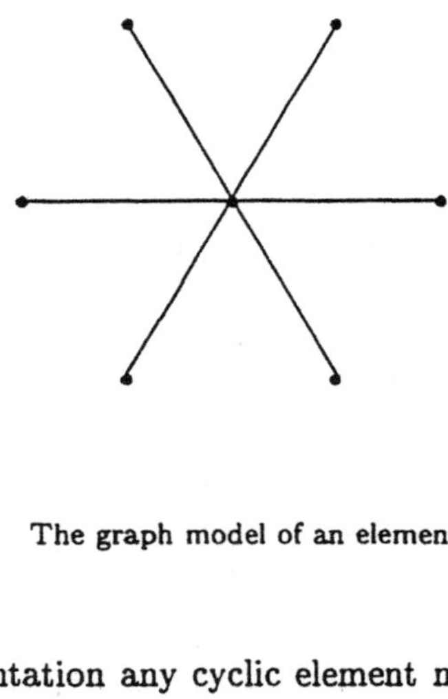

Fig. 4.3. The graph model of an element as a star

With such a representation any cyclic element may be placed in two positions (two orientations) which are obtained by mirroring images of each other. There exist some elements, however, which allow only one orientation. Unfortunately, there is no adequate graph model for the case.

When the order of pins is not critical we can delete all edges of the cycle from the representation obtained. This leads us to a star-shaped model of an element (Figure 4.3).

Now we shall consider the model of an element with duplicate pins. If these pins are neighbouring they can be represented as one vertex. Let there exist some pairs of duplicate pins. The model of an element with two pairs of duplicate pins $\{1,5\},\{8,12\}$ is shown in Figure 4.4. Figure 4.4 shows an

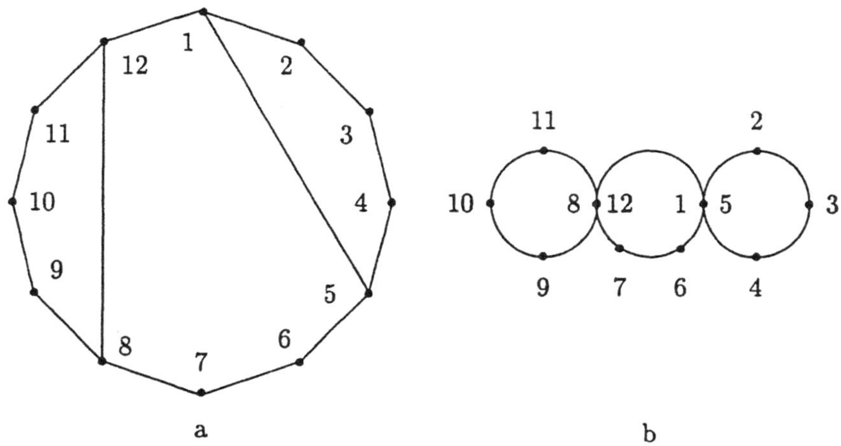

Fig. 4.4. (a) Elements with two pairs of duplicate pins: 1,5 and 8,12. (b) Graph model of the element as a 'cactus'.

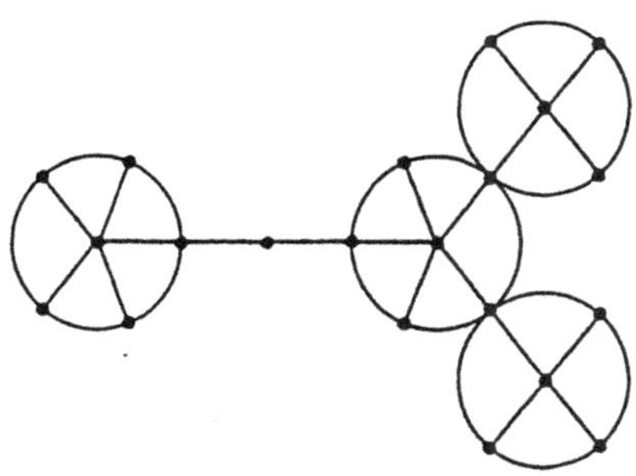

Fig. 4.5. The graph model of an element

adequate model of this element which has been obtained by contracting the vertices associated with the duplicate pins into one vertex. The described operation, generally, leads to a graph model called 'cactus' in graph theory. Leaves, trunk and branches of cactuses which correspond to element models consist of linear and cyclic components (Figure 4.5). Unfortunately, such models may be used only for a planar element, i.e. no routes between duplicate elements within the element cross each other. This model cannot be applied, then, to the element of Figure 4.6.

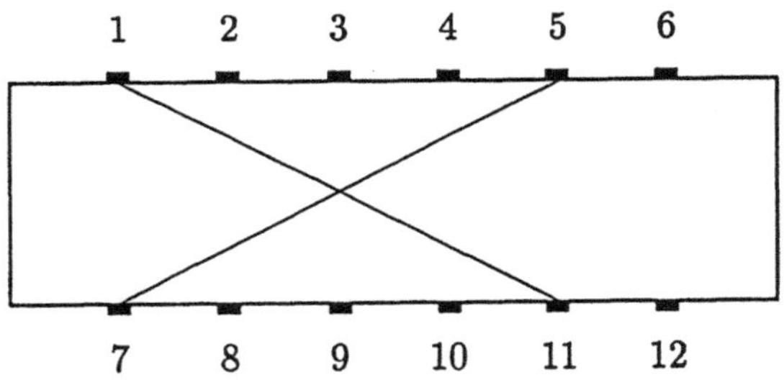

Fig. 4.6. An element with duplicate pins for which there is no cactus-form graph model

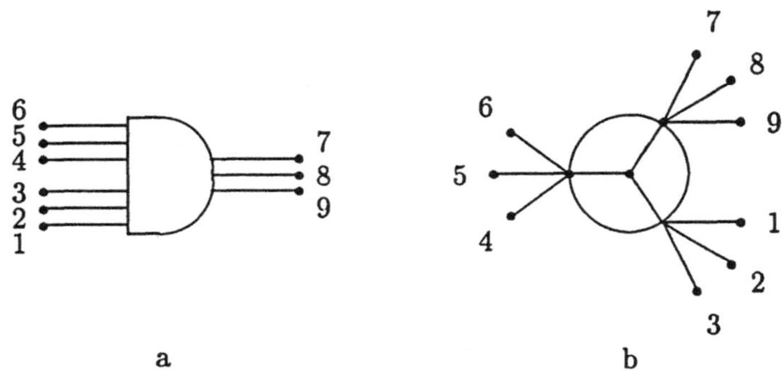

Fig. 4.7. (a) An element for which logically equivalent pins are arranged next to each other on the boundary: $1 \sim 2 \sim 3$, $4 \sim 5 \sim 6$, $7 \sim 8 \sim 9$. (b) The graph model of the element.

We suppose now that an element has logically equivalent pins arranged next to each other on the element boundary (Figure 4.7,a). Then we obtained the model of an element which is a cycle with stars representing the sets of logically equivalent pins threaded on it (Figure 4.7,b). Major problems might arise when logically equivalent pins are arranged arbitrarily on the element boundary. In this case, unfortunately, finding an an adequate graph representation for an element is highly improbable.

We can act in the same way for the logically equivalent groups of pins.

For more detail on graph models for elements refer to [62, 65, 66, 80, 81].

In [62] the authors proposed the model $\mathbf{M}_2$ of a circuit represented as a hypergraph obtained by adding a hyperedge formed by the vertices associated with the pins of the net to the union of graph models for the elements of each net in the circuit. Thus the vertex set of the hypergraph consists of the elements and central vertices. The hyperedge set consists of hyperedges which represent nets and edges forming the graph models of the elements (the so called *structural edges*). Since the graph models of elements applied in $\mathbf{M}_2$ do not completely account for logical and physical equivalence of pins nor does model $\mathbf{M}_2$.

Consider an updated version $\mathbf{M}_2'$ of the model described which allows us to take into account arbitrary arrangement of duplicate pins by rejecting the requirement of element planarity. For this purpose we shall represent each acyclic component as a cyclic one where each pair of opposite pins is duplicate. A central vertex of some elemant A_k and all vertices which correspond to pins of A_k duplicating v will be added to each hyperedge that corresponds to a net and contains the vertex v which is the pin of the element A_k.

A disadvantage of the model class in question, including the most general hypergraph model, is the inability to account completely for distribution of logically equivalent pin groups in the element as well as metric constraints for over-the-element routes. In particular, for over-the-element routes we propose either to introduce extra dummy vertices corresponding to channels involved in routing into their cyclic representations or to assign capacity values to structural edges, thereby indicating the number of interconnections allowable between the given pins in the element [62].

We shall use a hierarchical nongraph model for the elements to eliminate the above disadvantages. Planarization of the connection model $\mathbf{C}_1$ based on the analysis of element models reveals whether it is possible to implement the order of connection conveyance to the element that has been preselected at some step of planar embedding construction. Moreover, such a model allows us to find out whether over-the-element routes are possible. We shall discuss the hierarchical model for the element in more detail (Figure 4.8).

On the first (top) level in the hierarchy the set of pins in the element is classified as logically equivalent pin groups. Logical equivalence of pin groups $\mathbf{C}_1$ and $\mathbf{C}_2$ means that during circuit implementation we can replace all pins in $\mathbf{C}_1$ with the appropriate pins in $\mathbf{C}_2$ and all pins in $\mathbf{C}_2$ with the appropriate pins in $\mathbf{C}_1$, i.e. we can connect nets linked to pins in $\mathbf{C}_1$ to the respective pins in $\mathbf{C}_2$ and those linked to pins in $\mathbf{C}_2$ to the respective pins in $\mathbf{C}_1$.

Each class of logically equivalent pin groups consists of some pin groups which form the second level of the hierarchy. Each pin in its turn is divided into groups of logically equivalent pins. As before, logical equivalence of pins v_1 and v_2 means that if in the initial circuit model v_1 connects to net H_1 and

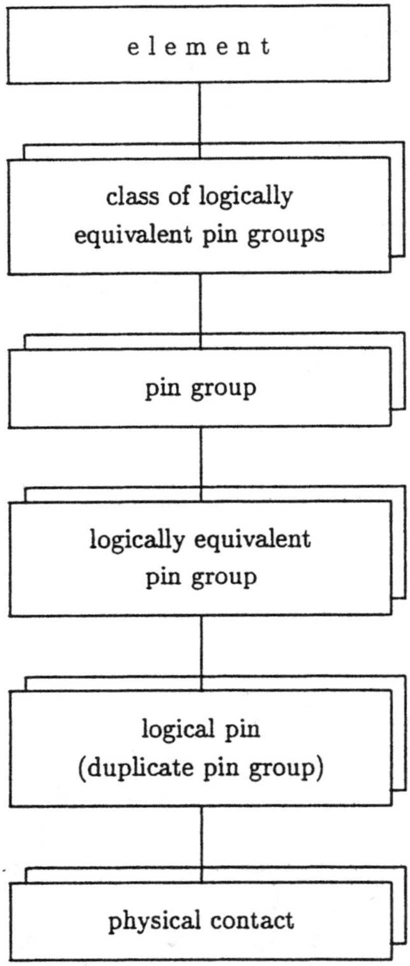

Fig. 4.8. The conceptual scheme of element hierarchical representation

v_2 connects to H_2 then permutation is allowed at circuit implementation, i.e. v_1 may connect to H_2 and v_2 to H_1. The groups of logically equivalent pins form the next, the third, level of hierarchy.

In its turn, each group of logically equivalent pins is broken down into groups of physically equivalent (duplicate), i.e. equipotential, pins. The groups of duplicate pins which will be henceforward called *logical pins*, form the forth level of hierarchy.

And, eventually, the fifth level of hierarchy forms physical pins. This level of hierarchy contains geometric information about pin arrangement on the boundary of an element, the maximum number of routes allowable between each pair of pins, etc. This information is used to find out whether the order of connection conveyance to an element is valid, whether an over-the-element route is feasible, etc.

Now we shall formulate a validity testing rule for the order of conveyance of edges to a vertex x of the model $\mathbf{C}_1$ for the circuit associated with an element A_x. First we assume that there are no equivalent pins among logical pins of A_x. Consider the cyclic order of edges $r_1, ...r_n$, $r_i = (x, v_i)$, $i = 1, ..., n$, which are incident with vertex x. Every edge r_i takes part in the implementation of some net H_i. Since by the assumption there are no equivalent logical pins then the points on the boundary of A_x which may participate in routing the connection are the duplicate pins from some set l_i. Thus each edge $r_i = (x, v_i)$ corresponds to a logical pin l_i, $i = 1, ..., n$, of element A_x. The cyclic order is valid if and only if for each l_i there exists a physical pin $c_i \in l_i$ such that the cyclic order of physical pins $c_1, ..., c_n$ coincides with the order of edges. In a general case, we consider a substitution s on the set of k integers where k coincides with the number of logical pins in the element. Substitution is called *pertinent* if any logical pin l_i is equivalent to $l_x(i)$. The cyclic order of edges $r_1, ..., r_n$ which are associated with logical pins $l_1, ..., l_n$ of the element in the original circuit model is pertinent if and only if there exists a physical pin c_i for some pertinent substitution s in each subset $l_x(i)$ such that the order of physical pins $c_1, ..., c_n$ coincides with this order of edges.

We consider now the problem of over-the-element routing. The geometric model of an element allows us to formulate it as a special case of the rectangular routing problem: given the pairs of points $(a_1, b_1), ..., (a_n, b_n)$ in the grid nodes on the boundary of the rectangle $ABCD$, we must connect point a_i with point b_i, $i = 1, ..., n$, along the grid lines or state that the route is illegal.

If a pair of points (a_1, b_1) is to be connected during construction of a plane embedding of a VLSI circuit the connection within the element (in the geometric model) is not routed. It is only fixed in the endpoints a_1, b_1. This allows us to expand the possibilites of over-the-element routing. Actually, the fixed connection between points a_1 and b_1 may prevent us from connecting points a_2 and b_2 although over-the-element routes are allowed for both of the pairs of points (a_1, b_1) and (a_2, b_2).

4.3. Dynamic VLSI Models

When a top-down layout design method is used layouts for each functional unit (cell) of VLSI are developed at each level starting from circuit schematics. Planarization of the i-th level cells results in establishing the order of connection routing to the *(i+1)*-th level cells. This order of routing along with the functional unit schematics specify the design requirements for layout of the *(i+1)*-th cells. Specifically, for each logical pin one must define many physically equivalent pins associated with it and their arrangement on the element's boundary.

From the above considerations it follows that, as opposed to the classical approach to the circuit planarization problem [62] of the preceding section the element models (represented by cells of the *(i+1)*-th level) are not prespecified but constructed dynamically during planarization. As a result we can speak about a dynamic VLSI model which extends from the VLSI layout draft on the top level represented by a functional diagram to the layout level.

Let hypergraph G be the model $\mathbf{C}_1$ of connections in the circuit. As noted above, the circuit planarity suggests the planarity of the hypergraph G, and the reverse is generally not true. However, the planarity of G for an i-th level cell suggests the planarity of layout for the cell. All nonplanar connections which might arise are transferred to layouts of next level cells because the obtained order of routing in these cells cannot be realized for planar cells. The said nonplanar connections will obviously exist, for example, if in a plane graph R there are two pairs of edges incident with the same vertex which belong to implementations of two distinct nets and intersect each other, i.e. when passing through the edges incident with some vertex (say in clockwise direction) we may encounter four edges e_1, e_2, e_3 and e_4 such that e_1 and e_3 participate in implementation of some net H_1 and e_2 and e_4 participate in implementation of net H_2 (Figure 4.9).

However, if intersections of nets within the elements in a circuit are not considered, the most adequate model of the circuit is $\mathbf{C}_1$, which generally is an arbitrary hypergraph.

Now we shall describe a special hypergraph model $\mathbf{C}_2$ which allows us to take account of the planarity of elements in the circuit. This model is naturally derived from a realization R of the model $\mathbf{C}_1$. For this purpose we shall substitute each vertex x in the graph R with a plane graph whose vertices are logical pins of the circuit's element which correspond to x. Now consider from where the edges of this subgraph originate. Let, for example, the vertex x of the graph R correspond to the element A_x of Figure 4.2. To this vertex

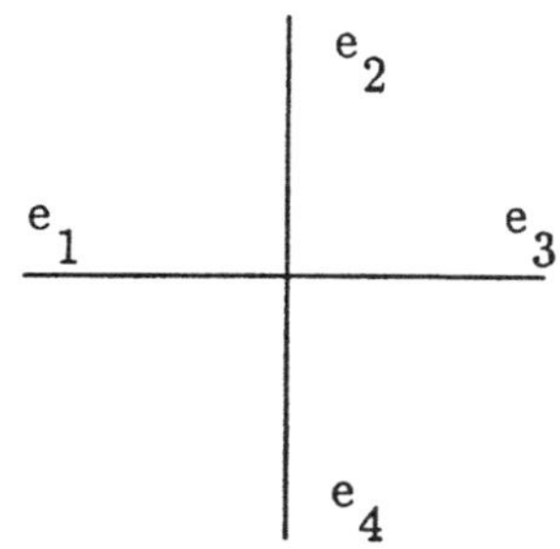

Fig. 4.9.

we assign a cycle whose vertices and order correspond to physical pins of A_x. Then all vertices associated with the duplicate pins are contracted into a single vertex. As we mentioned above, such pins are interconnected within the element. Due to the planarity of A_x the above operation results in a plane graph.

Alternatively, if the above described cross-wise configurations do not exist for a vertex x then the vertex can always be substituted by some tree such that the resulting graph remain planar, and the edges incident with the common vertex (excluding the tree edges) belong to realization of one and the same net.

In such a way we come to the model $\mathbf{C}_2$ of a circuit represented as a hypergraph G'. The hypergraph vertices are the pins of elements in the circuit and the hyperedges may be divided into two classes. The first class is formed by the hyperedges associated with the nets of the circuit. Each such hyperedge consists of all pins belonging to the same net. The second class is formed by hyperedges associated with the elements. Each hyperedge in this case consists of element's pins in the circuit. The hypergraph G' satisfies the following conditions:

- hypergraph vertices are of degree 2;
- all hypergraph cycles are of odd length.

We can readily demonstrate that an arbitrary hypergraph which satisfies the above conditions can be assigned to a circuit. As shown in section 3.2, the planarity testing problem is NP-complete for such hypergraphs.

Chapter 5

Extracting a Maximum Planar VLSI Part

5.1. Extracting a Maximum Planar VLSI Part as a Hypergraph Planarization Problem

In the preceding chapter we demonstrated that a hypergraph is generally a mathematical model of an electrical circuit. If nonplanar connections, i.e. over-the-element routes and those implemented by the connectors, are deleted from a circuit's implementation then we obtain the transformation

$$G = (V, \mathcal{H}) \rightarrow G' = (V, \mathcal{H}') \tag{5.1}$$

of the original hypergraph model to a planar hypergraph. This transformation is performed by splitting hyperedges, i.e. a hyperedge $H \in \mathcal{H}$ of the set $\mathcal{H}'$ is associated with the disjoint hyperedges whose union coincides with H.

If over-the-element routes are not allowed for VLSI embedding in a plane, the length of connectors becomes immaterial and the principal planarization criterion will be minimization of their number, the problem then consists in finding a transformation (5.1) with the minimal value of difference

$$t(G, G') = |\mathcal{H}'| - |\mathcal{H}|. \tag{5.2}$$

The problem of circuit planarity testing in a problem in its own right. When the model M_2 is used the problem reduces to testing the graph planarity with constraints on the orders of edges adjoining vertices. This constraint consists in the following: two structural edges (refer to item 4.2 for the model's definition) are called dependent if they are adjacent edges of some linear representation of an element. In a planar graph the pairs of dependent edges incident with a common vertex must not form a 'cross'. Figure 4.9 shows a 'cross' formed by two pairs of dependent edges (e_1, e_3) and (e_2, e_4). We shall call a planar graph which satisfies the above condition a Λ-*embedding*, by [62].

The Λ-embedding testing algorithm of [62] clearly has an exponential estimate for the running time. The question of whether polynomial algorithms exist for Λ-embedding testing in general remains open.

However, for a 3-connected graph, as stated in the well-known Whitney theorem (Theorem 2.3) about the uniqueness of its planar embedding, there exists a linear algorithm for Λ-embedding testing. Hence we can devise an algorithm whose running time is the sum of the linear function of the number of its vertices and the exponential function of the number of 3-connected components for an arbitrary graph.

Note that in the model $\mathbf{M}_1$ hyperedges are represented by vertices which form a single part of a bipartite graph. Transformation (4.1) consists in splitting the vertex of this part. Then an arbitrary bipartite graph can be considered as an $\mathbf{M}_1$ model for some circuit.

Proposition 5.1 *If connections can not be routed over element in a circuit the planarization problem for the model* $\mathbf{M}_1$ *with the minimum number of connectors belongs to the class of NP-complete problems.*

Proof. Let Γ be an arbitrary graph. On the graph Γ we construct a bipartite graph Γ_1 by splitting the edges of Γ , i.e. by replacing each edge $x = [v_i, v_j]$ with two adjacent edges $[v_i, u_x]$ and $[u_x, v_i]$ where u_x is a new vertex. We shall consider the constructed bipartite graph Γ_1 as an $\mathbf{M}_1$ model of a circuit where vertices u_x correspond to the nets. Then the splitting of such a vertex corresponds to removing the edge x from graph Γ. Hence the transformation (5.1), for which the minimum (5.2) was reached, uniquely determines the maximum planar subgraph of Γ. Since extraction of the maximum planar subgraph in an arbitrary graph is an NP-complete problem, the proposition is proved. □

The problem of minimization of the difference (5.2) for hypergraph models of circuits refers to the class of NP-hard. This follows from the NP-completeness of its subproblem which consists in testing a hypergraph for planarity. The latter is thus true (refer to item 3.2) both for arbitrary hypergraphs and for special-type hypergraphs with all vertices having degree 2 and all cycles of odd length.

Therefore the most interesting, from the practical standpoint, are approximation algorithms for solving the hypergraph planarization problem, i.e. constructing a planar hypergraph G' with the minimum difference value (5.2). There are many well-known approximation algorithms for solving the problems for graphs (refer to item 2.3). The majority of them ensure a solution which is at most thrice worse than the optimum one. At the same time the NP-completeness of the planarity testing problem for hypergraphs suggests the following.

Proposition 5.2 *If $P \neq NP$ then for any polynomial approximation algorithm for hypergraph planarization its relative performance ratio is ∞, i.e. for any integer K there exists a hypergraph for which the solution obtained by the algorithm is K times worse than the optimum one.*

The approximation algorithms for planarization discussed below are applicable for hypergraph planarization. Their specific application for VLSI planarization is primarily determined by the fact that some vertices in a hypergraph correspond to the elements for which pin sequence, allocation of logically and physically equivalent pins, etc. are specified. Application of the algorithms discussed below does not account for them. It should noted, however, that in a number of cases the use of different graph models for the elements provides for the required order of connection routing, but not necessarily always.

The decomposition planarization method discussed in item 3.4 allows us to reduce the solution of an arbitrary hypergraph planarization problem to that of similar problems for 'nondecomposable' hypergraphs. This chapter proposes two approaches to an approximate solution of such problems.

The first of them is an iteration method, each step of which determines and embeds some portion of a hypergraph (the so called segment). In this connection, the stated efficient criterion of segment embeddability becomes a key one. This method applied to the problem of graph planarity testing leads us to the well-known Demoucron-Malgrange-Pertuiset algorithm.

The second approach consists in constructing a hypergraph G' with the planar Konig representation obtained by splitting the vertices of the graph $K(G)$ corresponding to the edges of the hypergraph G on the initial hypergraph G. Generally, construction of the graph $K(G')$ by the minimal number of splits is an NP-complete problem. Below we propose an approximate algorithm for solving the problem based on Lempel-Even-Cederbaum's algorithm (refer to item 2.2) using PQ-trees. Planar realization of hypergraph G' can be either used as an initial embedding for the iteration algorithm which implements the first approach or extended to the maximal planar subgraph of some realization of the source hypergraph G by, for instance, a 'greedy' algorithm which inserts the edges joining the vertices from different parts of the split edges in hypergraph G.

We shall now discuss the algorithms in more detail.

5.2. Iteration Algorithm for Hypergraph Planarization

Let $G = (V, \mathcal{H})$ be an arbitrary connected hypergraph and $R_0 = (V_0, E_0)$ be a plane subgraph of some realization of G. If the subgraph $R_0(H)$ of the graph R_0 induced by the vertex set H is connected for any $H \in \mathcal{H}$, then R_0 is a planar realization of the hypergraph G. Otherwise, we shall denote by $\mathcal{H}_0$ the set of edges $H \in \mathcal{H}$ for which the subgraph $R_0(H)$ is unconnected.

The segment S of the hypergraph G with respect to R_0 is a partial hypergraph of G in one of the following forms:

1. the edge $H \in \mathcal{H}$ of the hypergraph G which is incident only with vertices from V_0;
2. the partial hypergraph induced by edges from $\mathcal{H}_0$ which are incident with the vertices from W, where W is the vertex set of a connected component of the subhypergraph of G induced by the vertex set $V \setminus V_0$.

The vertices of the segment which belong to the set V_0 are said to be *contact* vertices. An *external chain with respect to R_0* is the minimal chain in the segment which connects the contact vertices.

Now we shall consider, for instance, the hypergraph G of Figure 5.1. As R_0 we take the graph of Figure 5.2. There exist three segments of the hypergraph G with respect to R_0 (Figure 5.3):

$$S_1 = (\{1,3,7\}, \{\{1,3,7\}\}),$$

$$S_2 = (\{4,6,8,10\}, \{\{4,6,10\}, \{4,8,10\}\}),$$

$$S_3 = (\{7,11,12\}, \{\{11,12,3\}\}).$$

The contact vertices of segment S_1 are the vertices *1, 3* and *7*; those of S_2 are the vertices *4, 6* and *8*; and those of S_3 are the vertices *11, 12* and *3*. Segment S_1 contains a single external chain $\{1,3,7\}$, segment S_2 contains two chains: $\{4,6,10\}$ and $\{4,8,10\}$, and segment S_3 contains one chain: $\{3,11,12\}$.

Clearly, all segments of G with respect to R_0 can be determined by breadth-first or depth-first search in the Konig representation in time $O(q)$, where q is the number of edges in the Konig representation.

For an arbitrary hyperedge $H \in \mathcal{H}$, consider the partitioning of the set $H \cap V_0$ by connected components of graph $R_0(H \cap V_0)$:

$$H \cap V_0 = H_1 \cup \ldots \cup H_k.$$

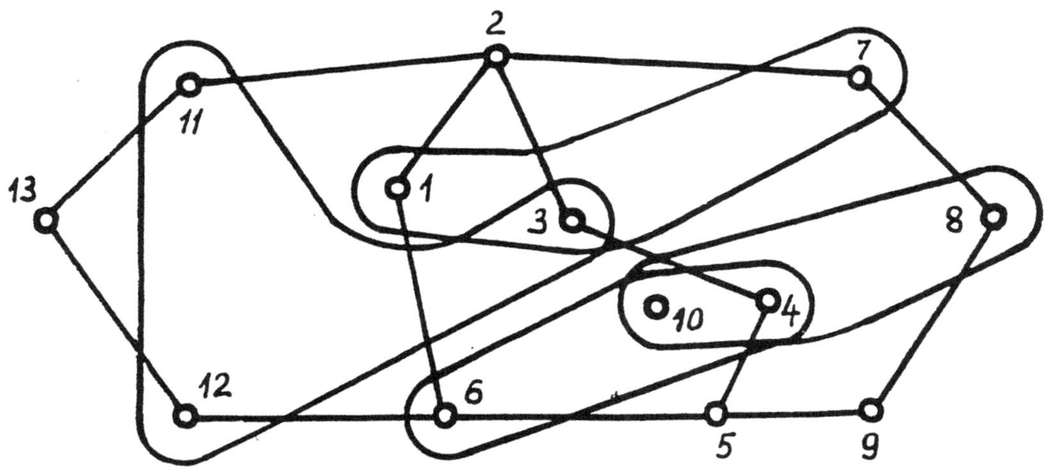

Fig. 5.1. An example of a hypergraph G.

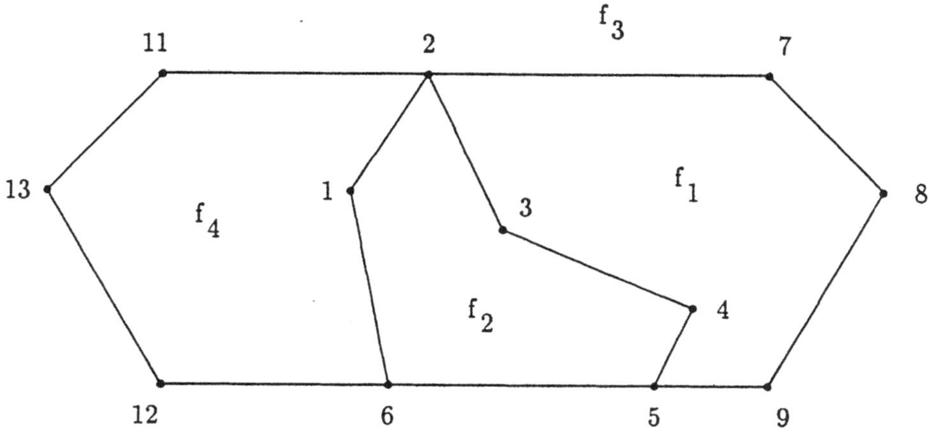

Fig. 5.2. The plane subgraph R_0 of some realization of the hypergraph G of Figure 5.1.

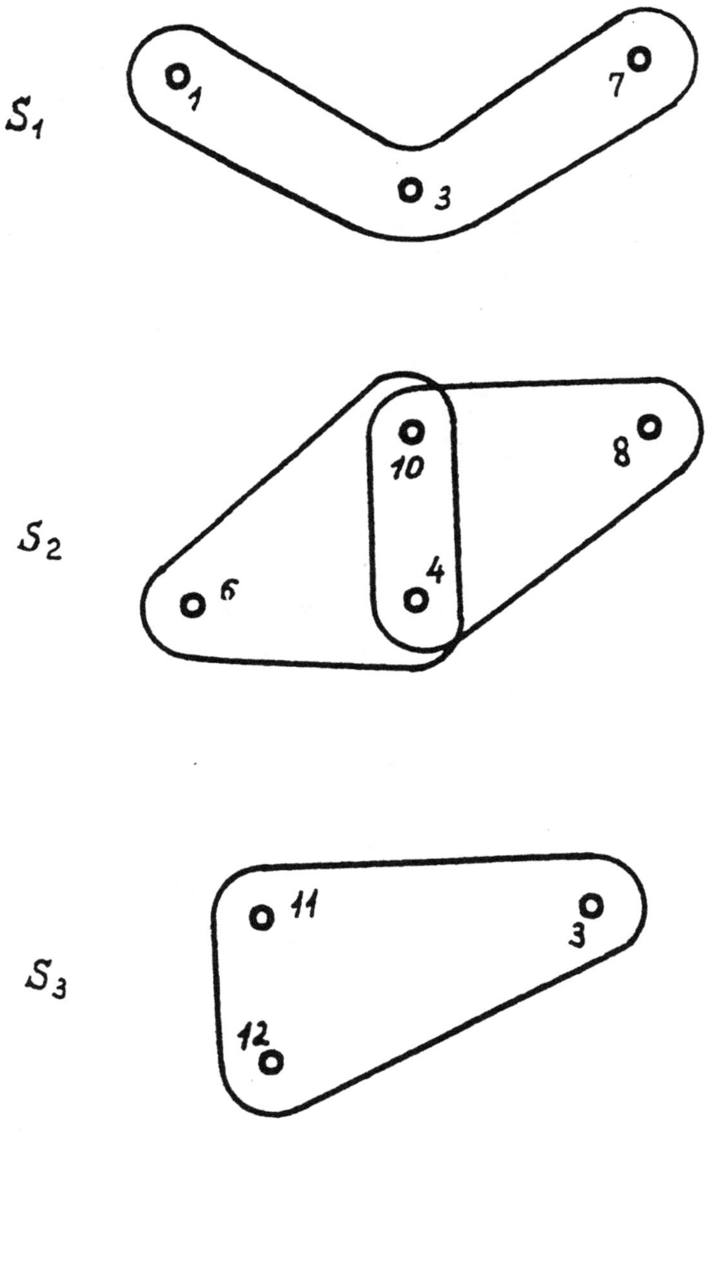

Fig. 5.3. Segments of the hypergraph G of Figure 5.1 with respect to the graph R_0 of Figure 5.2.

The subsets H_i in this partitioning we shall call *contact* subset. For segment S we define a bipartite multigraph $\Gamma(R_0, S)$ the first part of which consists of contact subsets of S and the second part consists of the faces of the planar embedding R_0 which contain the contact vertices of S. The pair $[H_i, f_j]$ forms an edge if there are vertices lying in f_j among the vertices of H_i, with the multiplicity equal to the number of such vertices. For the hypergraph G of Figure 5.1 and the subgraph R_0 of Figure 5.2, the subgraph $\Gamma(R_0, S_i)$, $i = 1, 2, 3$, are represented in Figure 5.4.

Proposition 5.3 *Let P be an external chain of the hypergraph G with respect to R_0. Then for the existence of a planar supergraph R of R_0 such that the subgraph $R(H)$ is connected for any edge of the chain P it is necessary and sufficient that all contact subsets of the chain P belong to the same connected component of the graph $\Gamma(R_0, S)$.*

The chain P which satisfies the requirement of proposition 4.2 we will call *embeddable* and the operation of constructing the subgraph R an *embedding of P*. Segment S is said to be *embeddable* if any pair of its contact subsets is connected by an embeddable chain.

Proposition 5.4 *A segment S of a hypergraph with respect to R_0 is embeddable if and only if the graph $\Gamma(R_0, S)$ is connected.*

Treating this proposition as an embedding extension criterion we can give a generalization of the Demoucron-Malgrange-Pertuiset iteration algorithm for hypergraphs.

Step 1. If an initial planar graph R_0 is not defined then we replace it with a cycle which is a realization of some cycle in the hypergraph G. Let $R = R_0$.

Step 2. Define segments $S_1, S_2, \ldots$ of the hypergraph G with respect to R. If no such segments exist, the algorithms stops running. R is a planar realization of hypergraph G.

Step 3. If the graph $\Gamma(R, S)$ is not connected for some segment S, the algorithm is completed: planar realization of the hypergraph G has not been found.

Step 4. If the graph $\Gamma(R, S)$ has an articulation vertex f corresponding to a face of R embedding for some segment S, we isolate an external chain

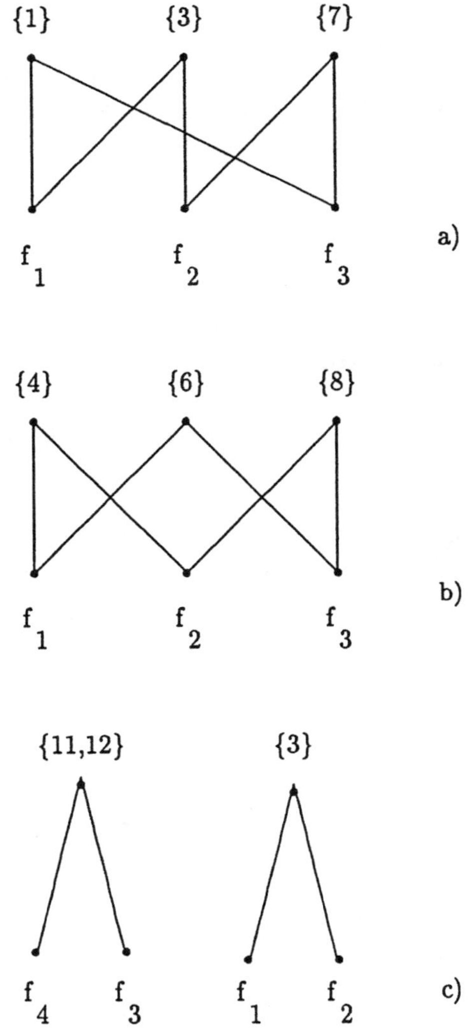

Fig. 5.4. a) The graph $\Gamma(R_0, S_1)$; b) the graph $\Gamma(R_0, S_2)$; c) the graph $\Gamma(R_0, S_3)$.

of the segment S which connects the contact subsets incident with the vertex f the same as vertices in $\Gamma(R, S)$. Embed this chain in face f and go to step 2.

Step 5. Select an arbitrary external chain, embed it and go to step 2.

Clearly, the presented algorithm for graphs coincides with [17].

We will now estimate the time complexity of the algorithm. Let n and m be the number of vertices and edges, respectively, in a hypergraph and q be the number of edges in its Konig representation. Then the time for constructing the segments (step 2) and checking the graphs $\Gamma(R,S)$ for connectedness (step 3) do not exceed $O(q)$, and the time complexity of step 4 does not exceed $O(n^2+q)$. Thus the time complexity of the algorithm does not exceed $O(mn^2+nq)$. Considering that q does not exceed $O(nm)$, we obtain a time complexity of $O(mn^2)$.

Now we shall discuss the application of the above mentioned algorithm for the approximated solution of the hypergraph planarization problem. This can be performed in two ways. The first one undeterministically selects a segment, an articulation vertex, and an external chain at steps 4 and 5. The second one makes deterministic selection by a heuristic procedure. Selection of different heuristic procedures generates a class of approximation algorithms for hypergraph planarization. Below we will discuss the first technique. As to the second one, we present an approach to constructing deterministic selection procedures.

If a segment S with an unconnected graph $\Gamma(R,S)$ has been detected at step 3 it clearly has two connected components. In that case we replace the original hypergraph G with a hypergraph G' by subdividing the segment S into two hypergraphs S' and $S"$ for which $\Gamma(R,S')$ and $\Gamma(R,S")$ are connected graphs. For instance, $\Gamma(R_0,S_2)$ is unconnected in the above example. We subdivided the segment S_2 into two hypergraphs $S_2' = (\{11,12\},\{\{11,12\}\})$ and $S_2" = (\{3\},\emptyset)$ for which graphs $\Gamma(R_0,S_2')$ and $\Gamma(R_0,S_2")$ are connected. Therefore we replace the original hypergraph G by the hypergraph G' obtained from G by removing the edge $\{11,12,3\}$ and adding the edge $\{11,12\}$. Such an operation can be carried out by constructing a minimal cut in the Konig representation in a time $O(qp^{2/3})$, where p and q are the numbers of vertices and edges, respectively, in the Konig representation.

The proposed deterministic procedure for selection of a segment, an articulation vertex, and an external chain consists in minimizing the number of segments for which the graphs $\Gamma(R,S)$ are unconnected. For this purpose an exhaustive search for a segment and an articulation vertex might be needed. An exhaustive search for pairs of contact subsets that are adjacent to the same face as the vertices of $\Gamma(R,S)$ and for external chains that connect these contact subsets leads to a significant increase in computational complexity. Thus the search for external chains leads to an algorithm with the exponential complexity estimate. Therefore we propose to determine external chains by depth-first search on a Konig representation of an appropriate segment rather than determinate their selection. Selection of contact subset pairs is determined by searching for the pairs whose vertices are adjacent in a planar embedding. Since the number of vertices and faces does not exceed $O(n)$ and the number of segments does not exceed $O(m)$, then the running time for the above-described deterministic selection procedure does not exceed $O(n^3m^2)$. Considering that the procedure is performed in each iteration, its application raises the time complexity of the algorithm to $O(n^3m^3)$.

5.3. Hypergraph Planarization by Konig Representation

Let Γ be a planar subgraph of the Konig representation $K(G)$ of a hypergraph G. For each vertex of Γ corresponding to a hyperedge of G, we select its incident edge with Γ and contract it. The resultant graph Γ^* is planar and each edge of it belongs to some hyperedge in the realization of G. Therefore, if graph Γ is a maximal planar subgraph of $K(G)$ then graph Γ^* can be taken as an approximated solution of the hypergraph planarization problem. To construct a maximal planar subgraph of $K(G)$, we can apply the well-known graph planarization algorithms that were described in Chapter 2, for instance.

Now we shall introduce a *vertex splitting operation*, i.e. replace a vertex v with vertices v_1 and v_2 and each edge $[u, v]$ with edge $[u, v_1]$ or $[u, v_2]$. Clearly, if the Konig representation $K(G)$ of the hypergraph G is nonplanar then by splitting vertices which correspond to edges of a hypergraph we can convert $K(G)$ into a planar graph Γ. As has been described earlier, we construct a planar graph Γ^* on the basic of the graph Γ. Each edge of Γ^* belongs to an edge in the hypergraph. Thus the problem of hypergraph G planarization 'is reduced' to constructing a planar graph Γ on the basis of a nonplanar graph $K(G)$ by the minimal number of splits on the vertices corresponding to the hypergraph edges. Now we shall discuss an approximation algorithm which permits us to construct a planar graph for an arbitrary nonplanar graph Γ by the minimal number of splits. The proposed algorithm is based on the well-known Lempel-Even-Cederbaum algorithm for graph planarity testing and its implementation by PQ-trees (item 2.2).

Let Γ be a 2-connected graph with the specified st-numbering. Similarly to item 2.2, we shall denote by Γ_i a subgraph of Γ induced by the vertex set $\{1, ..., i\}$ and let p be such that Γ_p is planar and graph Γ_{p+1} is nonplanar. The basic idea behind the algorithm is to construct a sequence of planar graphs $D_p = \Gamma_p, D_{p+1}, \ldots, D_n$. Each graph $D_k, k > p$, in this sequence has

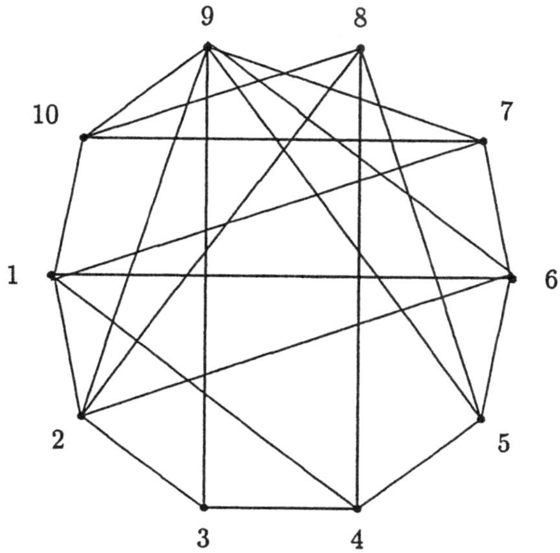

Fig. 5.5. The nonplanar *st*-graph Γ.

been obtained by a number of splits from Γ_k and contains graph D_{k-1} as an induced subgraph. In so doing we, naturally, want to minimize the number of vertices in graphs D_k.

Each vertex i of the graph Γ_k is associated with a collection of vertices $i_j, j \in J_{i,k}$ of graph D_k obtained by splitting the vertex i. Consider the graph Γ of Figure 5.5. For this graph $p = 8$ and the graph $D_8 = \Gamma_8$ is shown in Figure 5.6. Suppose we have already constructed graph D_9 (Figure 5.7). In D_9 vertex *9* of Γ_9 is associated with two vertices 9_1 and 9_2.

We denote by D_k' a graph obtained from D_k by adding for each edge $[i, r]$, $i \leq k$, $r > k$, of the graph Γ_k the edges $[i_j, r_{ij}]$, $j \in J_{i,k}$ where r_{ij} are new vertices labeled r. The graph D_k' for the example considered is shown in Figure 5.8.

Let us consider a plane embedding Y_k of the graph D_k for which all vertices labeled $k+1$ belong to the minimal number of faces. Each embedding Y_k of the kind is assigned a minimal number $g = g(Y_k)$ of groups $M_1, ..., M_g$ of consecutive vertices labeled $k+1$ such that for any edge $[i, k+1]$ of Γ_k in these groups there is a vertex r_{ij} for some $j \in J_{j,k}$. We now contract the vertices of every group M_l into a single vertex denoted by $(k+1)_l$ and delete the vertices

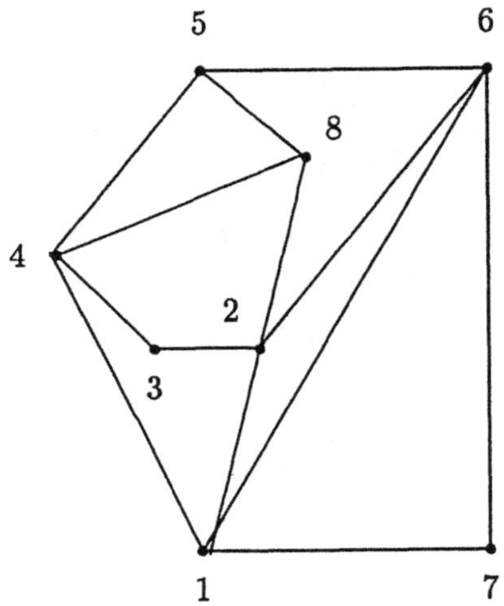

Fig. 5.6. The planar graph Γ_8

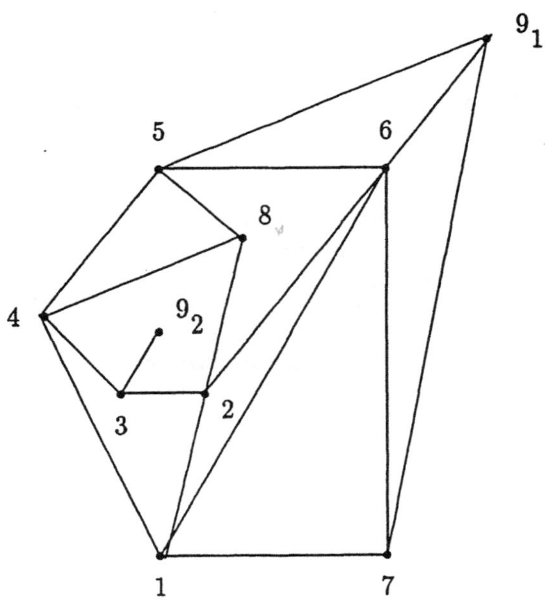

Fig. 5.7. The graph D_9

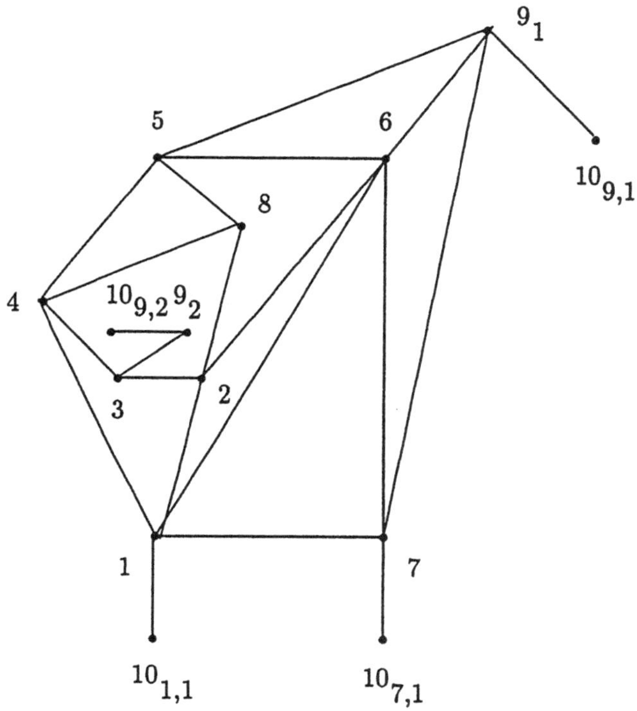

Fig. 5.8. The graph D_9'

labeled $k+1$ which belong to n sets of M_2. We denote the resultant graph by D_k''. Generally, an arbitrary edge $[i, k+1]$ of Γ_{k+1} is associated with a set $E_{i,k+1}$ of $[i_j, (k+1)_l]$-form edges in D_k''. Let us construct the graph D_{k+1} by retaining one edge from each set $E_{i,k+1}$ in the graph D_k''. Thus our problem is to find an embedding Y_k with the minimal value $g(Y_k)$ and construct the said collection of groups $M_1, ..., M_g$. To solve the problem we have applied the following approximation algorithm.

We now consider a collection $\{D_{kq}; q \in Q\}$ of all maximal subgraphs in D_k' obtained by removing some vertices labeled $k+1$ which are embeddable so that the remaining $(k+1)$-labeled vertices lie in the same face.

Let T_k^q be a PQ-tree corresponding to the graph D_{kq}' and S_{k+1}^q be a set of leaves in T_k^q that have been labeled by some edge from the set $E_{i,k+1}$ for an edge $[i, k+1]$ in G_{k+1}. Consider the minimal set $\{F_1^q, ..., F_t^q\}$ of vertex groups from S_{k+1}^q that are consecutively arranged. The number $t = t_q$ of such groups is said to be the size of the tree T_k^q. As an approximate solution of the problem, we take the smallest group set F_k from the set $\{F_j^q : q \in Q, j = 1, ..., t_q\}$ such that for any $[i, k+1]$ edge of graph Γ_k among the vertices of groups from F_k there exists a vertex labeled by an edge from $E_{i,k+1}$. The problem of searching for such a set F_k is a well-known NP-complete minimum cover problem [30]. Therefore, we shall confine ourselves to constructing a minimal set F_k only.

We shall now discuss the problem of calculating the size of the PQ-tree T_k. (In the following, the superscript q is omitted and we shall denote T_k^q simply T_k). Consider the following three types of PQ-trees. A PQ-tree is said to *belong to Type B* if vertices from S_{k+1} lie on both ends of its frontier. A PQ-tree *belongs to Type H* if some vertex from S_{k+1} lies on one end of its frontier only. The remaining PQ-trees will be assigned *Type A*, i.e. a PQ-tree *belongs to Type A* if the vertices which do not belong to S_{k+1} lie on both ends of its frontier.

Let v be an arbitrary vertex of the PQ-tree T_k. We denote by $T_k(v)$ a subtree of T_k induced by the vertices which are reachable from v and the vertex v itself. Determine three values $\mu_b(v)$, $\mu_h(v)$ and $\mu_a(v)$ that are equal to the minimal size of an equivalent PQ-tree $T_k(v)$ belonging to Type B, H or A, respectively. If none of the trees equivalent to $T_k(v)$ belongs to Type B, H or A, then $\mu_b(v) = \infty$, $\mu_h(v) = \infty$ or $\mu_a(v) = \infty$, respectively. Note that the values $\mu_b(v)$, $\mu_h(v)$ and $\mu_a(v)$ can be determined from the appropriate values for the children of the vertex v.

Thus by traversing down-up in tree T_k from leaves to the root we can find the values μ_b, μ_h, μ_a for each vertex of the tree. Having computed all these values for a pertinent root v_0, we assign Type B, H or A to tree $T_k(v_0)$ depending on which of $\mu_b(v_0), \mu_h(v_0)$ and $\mu_a(v_0)$ is the smaller. Then by traversal up-down from the pertinent root to leaves we assign Type B, H or A to each subtree $T_k(v)$ in accordance with the parent type. The information obtained allows us to construct an equivalent PQ-tree T_k of minimum size. Let us discuss the calculation of μ_b, μ_h and μ_a in more detail.

If all leaves of the tree T_k which are reachable from the vertex v belong to S_{k+1} then $\mu_b(v) = 1$, $\mu_h(v) = \infty$ and $\mu_a(v) = \infty$. If none of the vertices in S_{k+1} is reachable from v then $\mu_b(v) = \infty$, $\mu_h(v) = \infty$ and $\mu_a(v) = 0$.

Let v then be a P-vertex of the tree T_k, and let $v_1, ..., v_k$ be its children, $\mu_{ha} = \min\{\mu_h, \mu_a\}$, $\mu_{bh} = \min\{\mu_b, \mu_h\}$, $\mu_{bha} = \{\mu_b, \mu_h, \mu_a\}$. We denote by P_b the vertex set v_i, $i = 2, ..., s-1$ for which $\mu_{bha}(v_i) = \mu_b(v_i)$, by P_h the vertex set $v_i, i = 2, ..., s-1$ not belonging to P_b for which $\mu_{bha}(v)_i = \mu_h(v_i)$, and by P_a the vertex set v_i, $i = 2, ..., s-1$, not belonging to $P_b \cup P_h$ for which

$\mu_{bha}(v)_i = \mu_a(v_i)$. Let $n_b = |P_b|$, $n_h = |P_h|$, $n_a = |P_a|$ and

$$\varphi_b = \begin{cases} n_b - 1, & \text{if } n_b > 0, \\ 0, & \text{if } n_b = 0. \end{cases}$$

Calculate the value of $\mu_a(v)$. For this purpose we select two vertices with the minimum difference $\mu_{ha} - \mu_{bha}$ among the vertices $v_1, ..., v_k$. Without loss of generality, we may take that these are the vertices v_1 and v_k. If $\mu_{ha}(v_1) = \infty$ or $\mu_{ha}(v_k) = \infty$ then $\mu_0(v) = \infty$. Otherwise, consider the following three cases.

Case 1. $\mu_{ha}(v_1) = \mu_h(v_1)$ and $\mu_{ha}(v_2) = \mu_h(v_2)$.

$$\mu_a(v) = \begin{cases} \sum_{i=2}^{s-1} \mu_{bha}(v_i) - \varphi_b + \mu_h(v_1) + \mu_h(v_s) - [\frac{n_h}{2}] - 1, & \text{if } n_h > 0, \\ \sum_{i=2}^{s-1} \mu_{bha}(v_i) - \varphi_b + \mu_h(v_1) + \mu_h(v_s), & \text{if } n_h = 0, n_a > 0, \\ \sum_{i=2}^{s-1} \mu_{bha}(v_i) - \varphi_b + \mu_h(v_1) + \mu_h(v_s) - 1, & \text{if } n_h > 0, n_a = 0. \end{cases}$$

Case 2. $\mu_{ha}(v_1) = \mu_a(v_1) \neq \mu_h v_1$ and $\mu_{ha}(v_s) = \mu_h(v_s)$.

$$\mu_a(v) = \begin{cases} \sum_{i=2}^{s-1} \mu_{bha}(v_i) - \varphi_h + \mu_a(v_1) + \mu_h(v_s) - [\frac{n_h - 1}{2}] - 1, & \text{if } n_h \geq 1, \\ \sum_{i=2}^{s-1} \mu_{bha}(v_i) - \varphi_b + \mu_a(v_1) + \mu_h(v_s), & \text{if } n_h = 0 \end{cases}$$

Case 3. $\mu_{ha}(v_1) = \mu_a(v_1) \neq \mu_h v_1$ and $\mu_{ha}(v_s) = \mu_a(v_s) \neq \mu_h(v_s)$.

$$\mu_a(v) = \begin{cases} \sum_{i=2}^{s-1} \mu_{bha}(v_i) - \varphi_b + \mu_a(v_1) + \mu_h(v_s) - [\frac{n_h}{2}] - 1, & \text{if } n_h \geq 1, \\ \sum_{i=2}^{s-1} \mu_{bha}(v_i) - \varphi_b + \mu_a(v_1) + \mu_h(v_s), & \text{if } n_h = 0 \end{cases}$$

Calculate the value of $\mu_h(v)$. For this purpose we select two vertices with the minimum difference $\mu_{ha} - \mu_{bha}$ and $\mu_{bh} - \mu_{bha}$ among the vertices $v_1, ..., v_s$. Without loss of generality, we may take these vertices as v_1 and v_s. If $\mu_{ha}(v_1) = \infty$ or $\mu_{bh}(v_s) = \infty$, then $\mu_h(v) = \infty$. Otherwise, we shall consider the following three cases.

Case 1. $\mu_{ha}(v_1) = \mu_h(v_1)$ and $\mu_{bh}(v_s) = \mu_b(v_s)$.

$$\mu_h(v) = \begin{cases} \sum_{i=2}^{s-1} \mu_{bha}(v_i) - \varphi_b + \mu_h(v_1) + \mu_b(v_s) - [\frac{n_h}{2}] - 1, & \text{if } n_h > 0, \\ \sum_{i=2}^{s-1} \mu_{bha}(v_i) - \varphi_b + \mu_h(v_1) + \mu_b(v_s), & \text{if } n_h = 0, n_a > 0, \\ \sum_{i=2}^{s-1} \mu_{bha}(v_i) - \varphi_b + \mu_h(v_1) + \mu_b(v_s) - 1, & \text{if } n_h > 0, n_a = 0. \end{cases}$$

Case 2. $\mu_{ha}(v_1) = \mu_a(v_1) \neq \mu_h v_1$ and $\mu_{bh}(v_s) = \mu_b(v_s)$.

$$\mu_h(v) = \begin{cases} \sum_{i=2}^{s-1} \mu_{bha}(v_i) - \varphi_h + \mu_a(v_1) + \mu_b(v_s) - [\frac{n_h - 1}{2}] - 1, \text{ if } n_h \geq 1, \\ \sum_{i=2}^{s-1} \mu_{bha}(v_i) - \varphi_b + \mu_a(v_1) + \mu_b(v_s), \text{ if } n_h = 0 \end{cases}$$

Case 3. $\mu_{ha}(v_1) = \mu_a(v_1) \neq \mu_h v_1$ and $\mu_{bh}(v_s) = \mu_a(v_s) \neq \mu_b(v_s)$.

$$\mu_h(v) = \begin{cases} \sum_{i=2}^{s-1} \mu_{bha}(v_i) - \varphi_b + \mu_a(v_1) + \mu_h(v_s) - [\frac{n_h}{2}] - 1, \text{ if } n_h \geq 1, \\ \sum_{i=2}^{s-1} \mu_{bha}(v_i) - \varphi_b + \mu_a(v_1) + \mu_h(v_s), \text{ if } n_h = 0 \end{cases}$$

Calculate the value of $\mu_b(v)$. For this purpose we select two vertices with the minimum difference $\mu_{bh} - \mu_{bha}$. Without loss of generality, we may take these vertices as v_1 and v_s. If $\mu_{bh}(v_1) = \infty$ or $\mu_{bh}(v_s) = \infty$ then $\mu_b(v) = \infty$. Otherwise, we shall consider the following three cases.

Case 1. $\mu_{bh}(v_1) = \mu_b(v_1)$ and $\mu_{bh}(v_s) = \mu_b(v_s)$.

$$\mu_b(v) = \begin{cases} \sum_{i=2}^{s-1} \mu_{bha}(v_i) - \varphi_b + \mu_b(v_1) + \mu_b(v_s) - [\frac{n_h}{2}] - 1, \text{ if } n_h > 0, \\ \sum_{i=2}^{s-1} \mu_{bha}(v_i) - \varphi_b + \mu_b(v_1) + \mu_b(v_s), \text{ if } n_h = 0, n_a > 0, \\ \sum_{i=2}^{s-1} \mu_{bha}(v_i) - \varphi_b + \mu_b(v_1) + \mu_b(v_s) - 1, \text{ if } n_h > 0, n_a = 0. \end{cases}$$

Case 2. $\mu_{bh}(v_1) = \mu_h(v_1) \neq \mu_b v_1$ and $\mu_{bh}(v_s) = \mu_b(v_s)$.

$$\mu_b(v) = \begin{cases} \sum_{i=2}^{s-1} \mu_{bha}(v_i) - \varphi_h + \mu_h(v_1) + \mu_b(v_s) - [\frac{n_h - 1}{2}] - 1, \text{ if } n_h \geq 1, \\ \sum_{i=2}^{s-1} \mu_{bha}(v_i) - \varphi_b + \mu_h(v_1) + \mu_b(v_s), \text{ if } n_h = 0 \end{cases}$$

Case 3. $\mu_{bh}(v_1) = \mu_h(v_1) \neq \mu_b v_1$ and $\mu_{bh}(v_s) = \mu_h(v_s) \neq \mu_b(v_s)$.

$$\mu_b(v) = \begin{cases} \sum_{i=2}^{s-1} \mu_{bha}(v_i) - \varphi_b + \mu_a(v_1) + \mu_h(v_s) - [\frac{n_h}{2}] - 1, \text{ if } n_h \geq 1, \\ \sum_{i=2}^{s-1} \mu_{bha}(v_i) - \varphi_b + \mu_a(v_1) + \mu_h(v_s), \text{ if } n_h = 0 \end{cases}$$

Let v be a Q-vertex of the PQ-tree T_x and $v_1, ..., v_s$ be the descendants of v which appear in the same left-to-right sequence as in T_x. We determine a digraph $M(v) = (X, U)$ whose vertex set is $X = \bigcup_{i=1}^{s} X_i$ where $X_i =$

$\{v_i^b, v_{i1}^h, v_{i2}^h, v_i^a\}$, and the arc set is $U = \{(x,y) : x \in X_i, y \in X_{i+1}, i = 1,...,s-1\}$. Each vertex in $M(v)$ is weighted: vertex v_i^b has weight $\mu_b(v_i)$, vertices v_{i1}^h and v_{i1}^h have weight $\mu_h(v_i)$ and vertex v_i^a has weight $\mu_a(v_1), i = 1,...,s$. Arcs of the form $(v_i^b, v_{i+1,1}^h), (v_{i2}^h, v_{i+1,1}^h)$ and (v_i^b, v_{i+1}^b) have a weight of *1*, the remaining arcs have a weight of *0*. From the constructions we can see that $\mu_a(v)$ is equal to the shortest path length connecting a vertex from the set $\{v_1^a, v_{12}^h\}$ with a vertex from $\{v_s^a, v_{s1}^h\}$; $\mu_h(v)$ equals the shortest path length connecting a vertex from the set $\{v_1^b, v_{11}^h\}$ with a vertex from $\{v_s^b, v_{s2}^h\}$; $\mu_h(v)$ equals to the minimum of the two numbers, the shortest path length connecting a vertex in $\{v_1^a, v_{12}^h\}$ with a vertex in $\{v_s^b, v_{s2}^h\}$ and the shortest path length connecting a vertex in $\{v_1^b, v_{11}^h\}$ with a vertex in $\{v_s^a, v_{s1}^h\}$. The time complexity for such a shortest path algorithm does not exceed $O(s)$.

We shall now estimate the time complexity for the algorithm proposed. Since the number of vertices in the graph D_i, $i = 2$, does not exceed n^2, then the time complexity of constructing a PQ-tree T_i' equivalent to T_i with the minimum size does not exceed $O(n^2)$. The time complexity for a minimal set F does not exceed $O(n^2)$. Thus the time complexity of the developed algorithm is $O(n^3)$.

It is interesting to note that the planarization of the hypergraph G leads to splitting only those vertices in the Konig representation $K(G)$ that correspond to hypergraph edges. Therefore, we shall naturally modify the above-discussed algorithm.

Suppose the vertex $k+1$ in the graph $K(G)$ corresponds to a vertex in the hypergraph G and it has been split into $t \geq 2$ vertices v_l, $l = 1,...,t$, at the k-th step of the algorithm. Among the vertices v_l we shall select a vertex v_i with the maximum degree. Then we remove all vertices v_l distinct from v_i from the graph B_{k+1} and insert an edge $[v_i, v]$, where v is a new vertex for each removed edge $[v_l, v]$. The resultant graph may be treated as D_{k+1}.

5.4. Operations of Local Embedding Optimization

At one time the investigators of approximate algorithms were concerned with finding a relationship between the solution obtained by the algorithms and the optimum solution. Another important property of an approximate solution for an optimization problem is its maximality, i.e. the absence of another 'largest' solution with the 'best' value of an objective function. For example, a maximal planar subgraph is the subgraph which becomes nonplanar by inserting an edge. Generally, a 'maximal' solution can significantly differ from the 'maximum' one, but nevertheless maximality is a most important property of an approximation algorithm.

Unfortunately, the algorithms discussed in Chapter 5 do not allow us to extract a maximal planar part neither in a hypergraph nor in a graph. The latter, however, can be readily corrected (for a time $O(nm)$ where n and m are the numbers of vertices and edges in a hypergraph, respectively). For this purpose it would be sufficient to scan the edges which do not belong to the extracted subgraph and insert such edges which do not make the subgraph nonplanar. As to constructing a maximal planar part of a hypergraph, this problem is NP-hard.

The above considerations imply that there exist embedding optimization operations which allow us to extend an earlier obtained embedding.

The first group is formed by operations which reconfigure realizations of hyperedges. Suppose that an edge e_1 can be drawn such that it crosses an edge e_2. Removal of the edge e_2 causes a hyperedge realization in the original hypergraph to be divided into two connected components H_1 and H_2. If after removal of the edge e_2 and insertion of the edge e_1 an embedding contains vertices x and y from H_1 and H_2, respectively, which belong to the same face, then e_2 may be replaced with the edge $[x, y]$. Clearly, the said operation allows us not only to extend the embedding but also decrease the number of intersections at constructing nonplanar connections in future (Figure 5.9).

Another group is formed by the following operations:

- rotation of biconnected components (Figure 5.10);
- changing a face of a 3-connected component embedding (Figure 5.11).

Local optimization operations of the third form lead us to varying the planar part rather than its extension and consist in finding an edge set $\{e_1, ..., e_k\}$ whose deletion makes it possible to embed a greater number of edges (Figure 5.12). Since the time complexity of such an operation is high it should naturally be performed for a relatively small k.

The above discussed local optimization operations should be applied until at least one operation allowing us to improve the embedding is reached.

It should be noted, however, that the problems of local embedding optimization, selection of sufficiently complete operation set, and their formalization, call for further in-depth investigation.

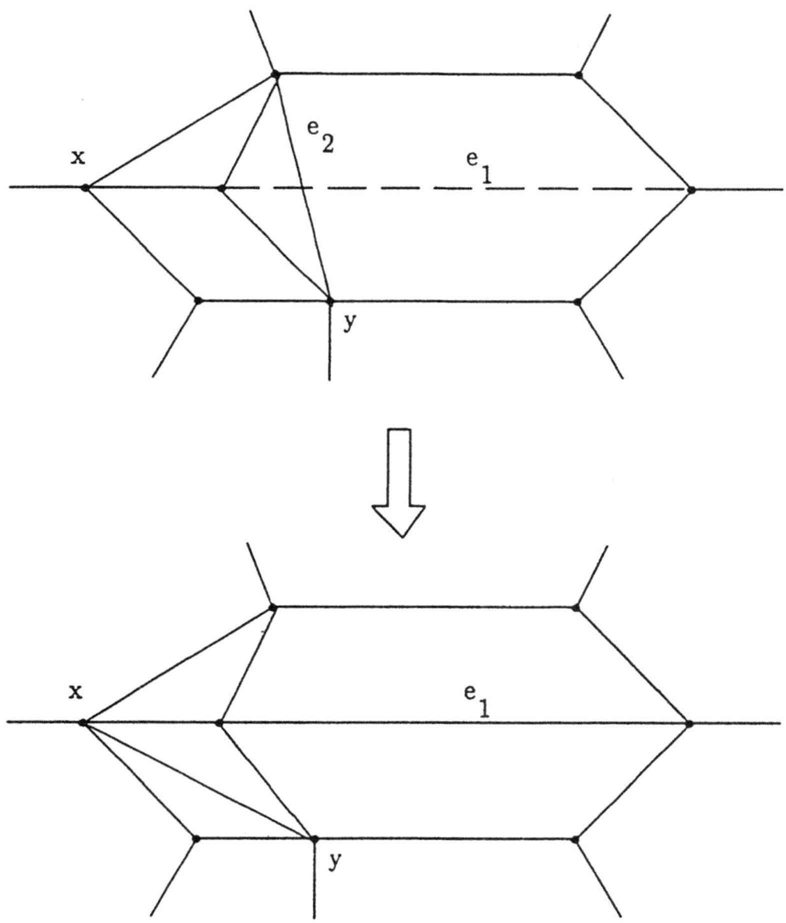

Fig. 5.9. Embedding optimization by applying the local operation of the first group.

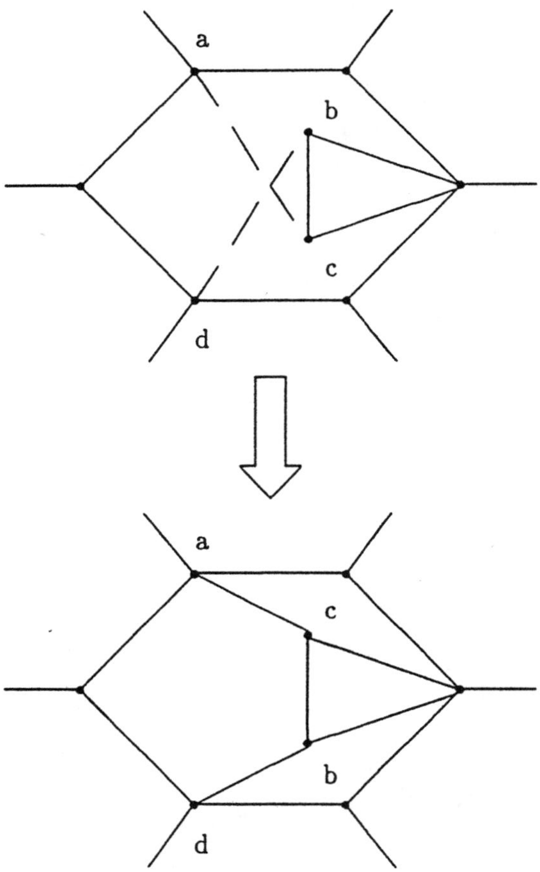

Fig. 5.10. Embedding optimization by rotating a biconnected component.

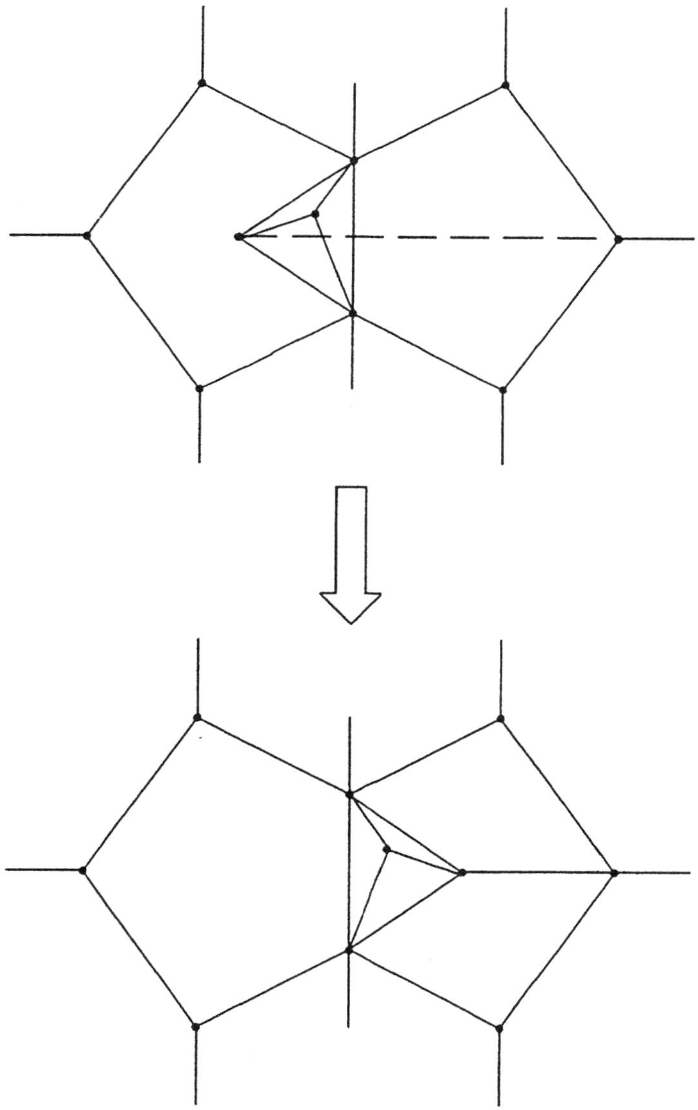

Fig. 5.11. Embedding optimization by varying the face for embedding a triconnected component.

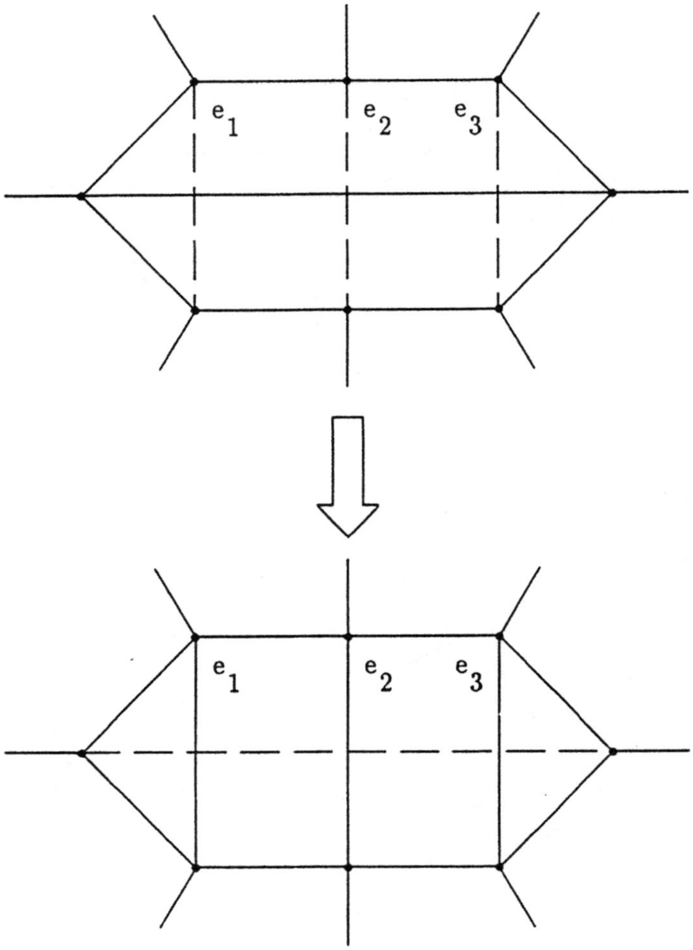

Fig. 5.12. Embedding optimization by applying the local operation of the third group.

Chapter 6

Constructing Nonplanar Connections

To construct a connection that has not been embedded at the first step of planarization for an electric circuit, one can use both over-the-element routes and connectors (diffused tunnels, jumpers, etc.). The special nonplanar connection graph described below may serve as a mathematical model for solving the problem.

6.1. Graph Model for Nonplanar Connections

Let $R(V, E)$ be a planar realization of some planar part of the circuit and $X \subseteq V \cup E$ be an identified subset of vertices and edges in the graph R. A *nonplanar connection graph* for embedding R related to the set X is a graph $N(R, X) = (F, Y)$ whose vertex set coincides with the face set of R. Two vertices f_i and f_j are adjacent in $N(R, X)$ if

1. either faces f_i and f_j are incident with a vertex from X, or
2. there exists a sequence

$$f_i = f^{(1)}, e_1, f^{(2)}, e_2, ..., f^{(n)} = f_j$$

of faces in R and edges in X such that

a) an edge e_k is incident with faces $f^{(k)}$ and $f^{(k+1)}$ for $k = 1, ..., n-1$;

b) all edges $e_1, ..., e_{n-1}$ correspond to one and the same net in the circuit.

The subset X is of crucial importance in the determination of the graph $N(R, X)$. It virtually determines the circuit elements and the already routed connections which can be used for constructing nonplanar connections. Thus, for example, if $X = V$ then over-the-element routes can be applied and connectors are forbidden. Alternatively, if $X = E$ then over-the-element routes are forbidden and only connectors can be placed. If $X = V \cup E$ then both connectors and over-the-element routes may be applied.

Suppose that extraction of a maximal planar part in a circuit caused a net H to be divided into two sections H_1 and H_2, i.e. the subgraph $R(H)$ of graph R induced by the vertex set H consists of two connected components with the vertex sets H_1 and H_2. Now we define the set $F(H)$, $i = 1, 2$, of the faces in R which may be used to bring the net H to section H_i. Then the construction of inter connection between sections H_1 and H_2 corresponds to a chain in $N(R, X)$ which joins a vertex in $F(H_1)$ with a vertex in $F(H_2)$.

Proposition 6.1 *Interconnection between sections H_1 and H_2 can be routed if and only if the vertices from the sets $F(H_1)$ and $F(H_2)$ belong to some connected component of the graph $N(R, X)$.*

The graph $N(R, X)$ is generally weighted. In fact, we must distinguish between the 'cost' of an over-the-element route and placement of a connector. The 'cost' of an over-the-element route can differ significantly depending on the element. For example, in a bipolar circuit, the 'cost' of inter connection between base and collector in a transistor is much higher than that of connection through a passive element.

Proposition 6.2 *The minimum 'cost' of connection routing between two sections H_1 and H_2 is equal to the length of the shortest chain in $N(R, X)$ which joins a vertex from $F(H_1)$ and a vertex from $F(H_2)$.*

Suppose the chain P connecting the sets $F(H_1)$ and $F(H_2)$ exists and has been defined. The routing of connections along this chain leads to correction of the graph $N(R, X)$. First, this may be attributed to the fact that new faces appear in the embedding. Second, connection of faces f_1 and f_2 routed over element x can lead to the fact that other faces incident to x cannot be connected over x. This is explained by the fact that either the connection between these faces intersects that between f_1 and f_2 or there are metric constraints for element x.

Determination of edges in graph $N(R, X)$ after routing some over-the-element connections in the circuit reduces to the following problem.

Let $(a_1, b_1), \ldots, (a_n, b_n)$ be points located in grid intersection nodes specified on the boundary of a rectangle. We want to connect point a_i with point b_i, $i = 1, \ldots, n$, along the grid lines or to know whether such a connection is possible. Here we naturally assume that connections between the point pairs (a_i, b_i) do not intersect.

A point pair (a_i, b_i) is said to be a *minimum* if by moving clockwise along the rectangle boundary from a_i to b_i no points a_j, b_j, $j \neq i$, are encountered.

Proposition 6.3 *If there exists a connection between pairs* $(a_1, b_1), ..., (a_n, b_n)$ *then the minimum pair can be found among them.*

From the proposition follows the algorithm for solving the problem stated above.

Step 1. Let $M = \{(a_i, b_i) : i = 1, ..., n\}$.

Step 2. Find the minimum pair (a_i, b_i) in M. If no such pair exists it is impossible to connect all pairs, and the algorithm terminates. Otherwise, go to step 3.

Step 3. Apply the minimum pair connection routing algorithm (given below) to pair (a_i, b_i). If no connection has been routed then connection of all pairs is impossible and the algorithm terminates. Otherwise, go to step 4.

Step 4. If (a_i, b_i) is a unique element of set M then all pairs $\{(a_i, b_i)\}_{i=1}^{n}$ have been connected and the algorithm completes. Otherwise, assume $M = M \backslash \{(a_i, b_i)\}$ and go to step 2.

Consider an algorithm for routing the minimum pair (a_i, b_i) connection. Source data for the algorithm is a rectangle with connections for a collection of point pairs (a_i, b_i). To construct a connection for the pair (a_i, b_i) we can use a line-expansion type of algorithm. In so doing, movement from the current point a^k proceeds as follows. The direction from point a^k is scanned in decreasing priority (Figure 6.1) to find the first point a^{k+1} such that a^{k+1} is a grid node adjacent to a^k and either $a^{k+1} = b_i$ or a^{k+1} is not an obstruction. In this case obstructions are the points on the rectangle boundary, points of already constructed connections and special deadlock points which will be detected by the below algorithm.

Step 1. Let $k = 0$, $a^k = a_i$.

Step 2. If the point a^{k-1} cannot be selected by the above rule (i.e. all grid nodes adjacent to a^k are obstructions) then go to step 3. If point a^{k+1} has been selected, go to step 4.

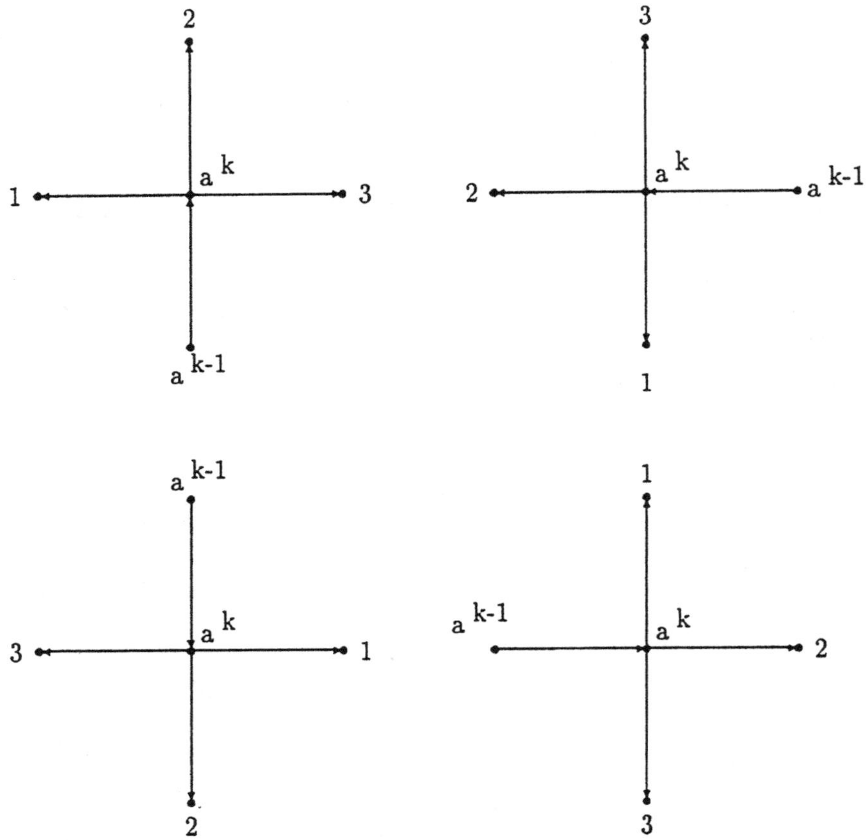

Fig. 6.1. Priority of directions at traversal from point a^k depending on the direction $a^{k-1} \to a^k$.

Step 3. If $a^k = a_i$, no connection is possible. The algorithm completes. If $a^k \neq a_i$ then mark a^k as a deadlock point, let $a^k = a^{k-1}$, $k = k+1$, and go to step 2.

Step 4. If $a^{k+1} = b^i$ connection is constructed and the algorithm completes. If $a^{k+1} = b_i$ then let $a^k = a^{k+1}$, $k = k+1$ and go to step 2.

Connecting sections H_1 and H_2 of the net leads to correction of the planar embedding obtained by extracting the maximum planar part. If the connection has been routed over the element associated with vertex x of the planar

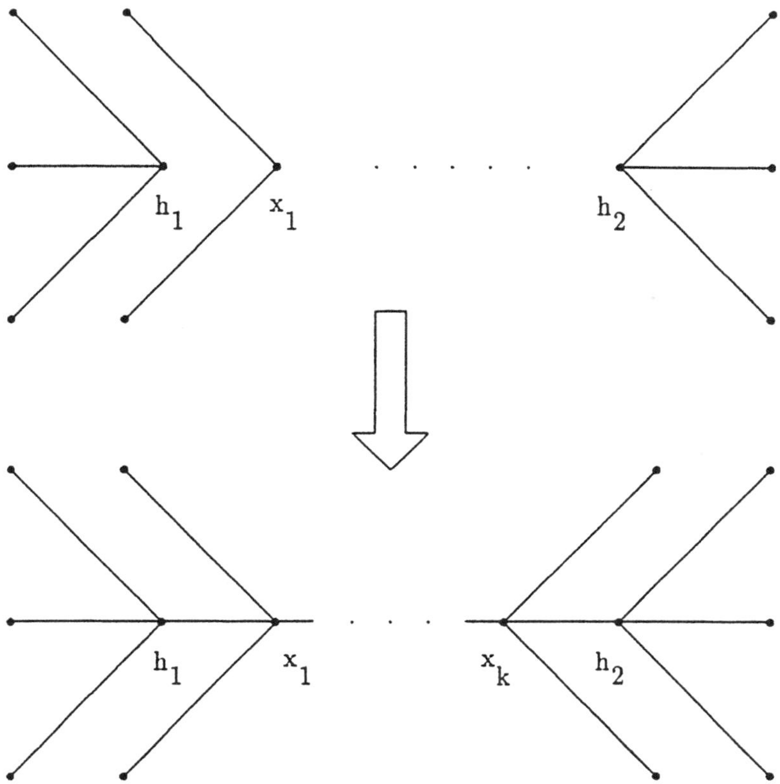

Fig. 6.2. Planar embedding transformation at connecting the net sections by over-the-element route.

embedding then edges $[h_1, x]$, $[h_2, x]$, where h_i is some vertex from H_i, $i = 1, 2$, are added to the planar embedding. Connecting the net sections H_1 and H_2 by routing over some elements associated with vertices $x_1, ..., x_k$ causes the edges $[h_1, x_1]$, $[x_1, x_2]$,..., $[x_k, h_2]$, where h_i as before is some vertex from H_i, $i = 1, 2$, to be added to the planar embedding (Figure 6.2).

Now let sections H_1 and H_2 of the net be connected by routing connections across edges

$$e_{i1}, ..., e_{i2_i}, ..., e_{k1}, ..., e_{k2_k}$$

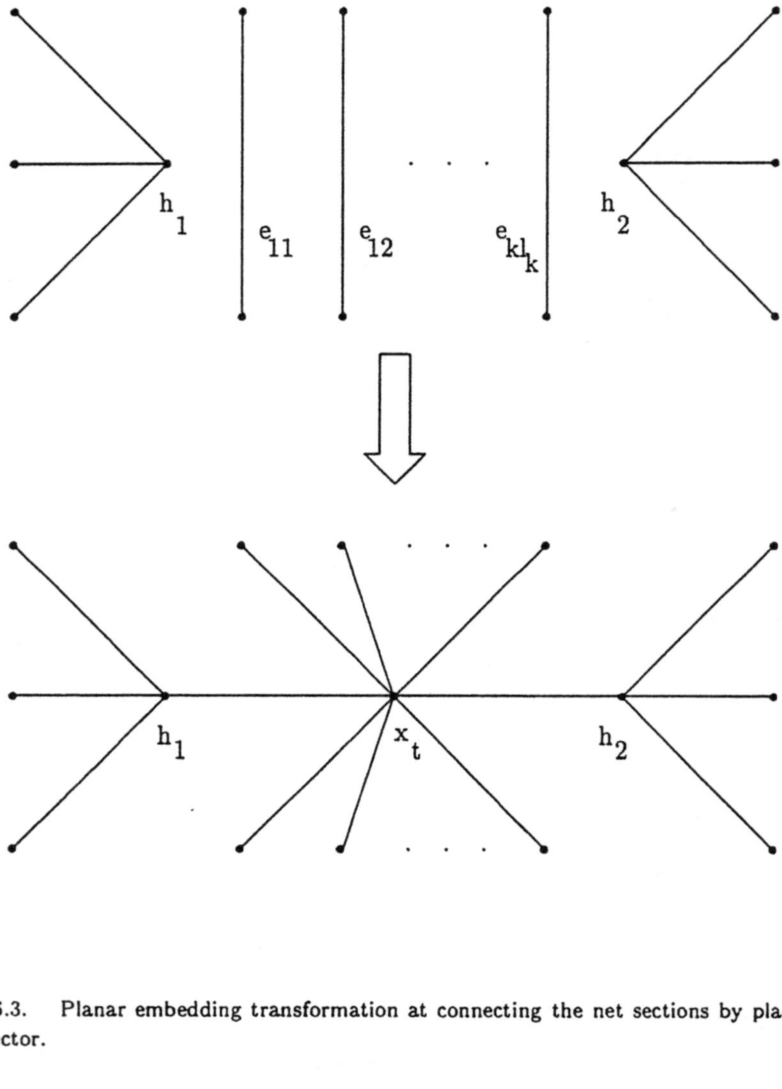

Fig. 6.3. Planar embedding transformation at connecting the net sections by placing a connector.

of a planar embedding with edges $e_{i1},...,e_{i2_i}$ associated with some net $H^{(i)}$. In this case the circuit is added with a connector which connects the sections H_1 and H_2. Nets $H^{(i)}$ are divided into sections connected through this connector. The planar embedding is added with a new vertex x_i, all edges $e_{ij}i$, $j = 1,...,l_i$, $i = 1,...,k$, are deleted and edges connecting the obtained sections of $H^{(i)}$ through the connector and edges $[h_1, x_i]$, $[x_i, h_2]$, where h_i is as before some vertex from H_i are added (Figure 6.3).

6.2. Constructing a Nonplanar Connection as a Steiner Problem in a Weighted Graph

Consider a net H which is divided into several sections

$$H = H_1 \cup ... \cup H_k \tag{6.1}$$

in the embedding R of the planar part. To construct a nonplanar connection between the sections of H_i we shall use the nonplanar connection graph $N(R, X)$ that has been introduced in the preceding item. In this case, the problem of constructing such a connection with a minimum total length of connectors and the minimum number of over-the-element routes in the circuit is reduced to the problem of constructing the minimum (by weight) tree with s pendant vertices from the set $F(H_1), ..., F(H_k)$, respectively, in the graph $N(R, X)$. Such a tree is called the *Steiner tree* which connects the vertex set $F(H_i)$, $i = 1, ..., s$.

If $|F(H_i)| = 1$ for all $i = 1, ..., s$ then we obtain the classical Steiner problem in a graph (refer to item 1.4). Recall the statement of the problem. Given graph $G = (V, E)$ whose edges are assigned non-negative weights, and $S \subseteq V$ is the vertex subset from V, $|S| = s$. We want to find a minimum weight subtree T whose vertex set contains S.

There are many papers devoted to the problem (refer, for example, to the survey [88]). Note that for any fixed S the Steiner problem is solvable in polynomial time. Generally, it remains NP-hard even in some special cases [88]. Nevertheless it is easy enough to develop an approximation algorithm with performance ratio at most 2, i.e. an algorithm for which the relation of the resultant tree weight C to the Steiner tree weight C^* is at most 2:

$$\frac{C}{C^*} \leq 2.$$

Below we discuss two of such algorithms which are more simple and 'natural'.

We denote by $PATH(W, v)$ the shortest chain connecting vertex v with vertices from W and by $C(W, v)$ the weight of the chain.

Algorithm 1 [44]

The algorithm starts operation from the subgraph $T_1 = (V_1, E_1)$, where $V_1 = \{v_1\} \subseteq S$ and $E_1 = \emptyset$. At each successive step we construct the graph T_i for $i = 2, 3, ..., k$ by adding to T_{i-1} the $PATH(V_{i-1}, v_i)$ where

$$c(V_{i-1}, v_i) = \min\{c(V_{i-1}, v_j) : v_j \in S \backslash V_{i-1}\}.$$

The tree T_k is an approximate solution for the Steiner problem. The time complexity of the algorithm does not exceed $O(sn^2)$, where $n = |V|$.

Algorithm 2 [76]

Step 1. Construct the complete graph $G_1 = (V_1, E_1)$, where $V_1 = S$. Each edge $[v_i, v_j] \in E_1$ is assigned a weight which is equal to the length of the shortest chain in G joining v_i and v_j.

Step 2. Find the minimum spanning tree G_2 of G_i.

Step 3. Construct a subgraph G_2 of G by replacing each edge of G_2 with an appropriate shortest chain of graph G. (If there is more than one shortest chain any may be selected).

Step 4. Find the minimum spanning tree G_4 in G_3.

Step 5. On the graph G_4 construct the Steiner tree by removing, if necessary, the edges such that all leaves belong to the set S.

The algorithms discussed above for solving the Steiner problem can be modified for constructing a Steiner tree which generally connects the sets $F(H_1)$. For this purpose the shortest chains between vertex subsets rather than vertices must be considered. Specifically, algorithm 2 may be modified as follows. For each pair of sets $F(H_i)$ and $F(H_j)$ we define the shortest chain P_{ij} connecting them. Among all these chains we select a minimum weight chain $P_{i_0j_0}$, and in partition (6.1) we merge the sections H_{i_0} and H_{j_0} and unite the appropriate subsets $F(H_{i_0})$ and $F(H_{j_0})$. The procedure repeats as long as the number of subsets in the partition (5.1) is greater than one. The union of all thereby obtained chains $P_{i_0j_0}$ will be an approximate solution for the problem in question.

Having constructed the spanning tree, we must construct over-the-element routes. In so doing, it may turn out that no extra connection for a pair of points (a_{n+1}, b_{n+1}) can be added to the already existing connections for point pairs $(a_1, b_1), ..., (a_n, b_n)$ although by applying the connection routing algorithm to the sequence $(a_1, b_1), ..., (a_{n+1}, b_{n+1})$ we have connected all point pairs (Figure 6.4). Therefore, we recommend appling the above algorithm for

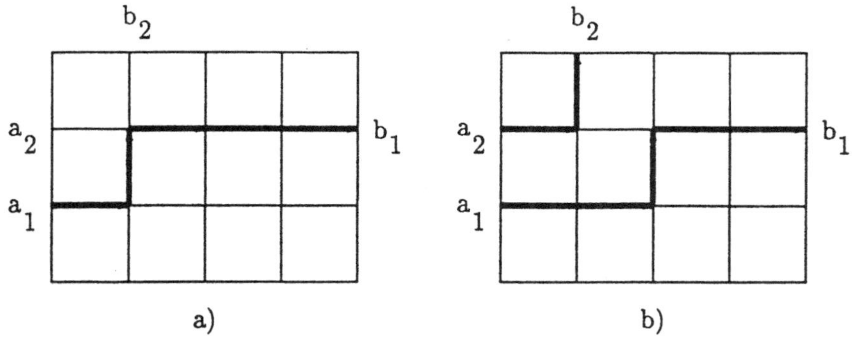

Fig. 6.4. (a) Fixed connection of pair (a_1, b_1) makes impossible the connections of the pair (a_2, b_2). (b) Implementation of connection for pairs (a_1, b_1) and (a_2, b_2).

the required connections. When the algorithm completes, it is necessary to minimize the length of wires for the obtained connections. Such optimization can be reached by a maze running algorithm as follows.

Step 1. Arrange pairs $(a_i, b_i), i = 1, ..., n$, with respect to the condition $(a_i, b_i) \leq (a_j, b_j)$ if a connection between a_j and b_j has been constructed earlier than that between a_i and b_i during execution of the algorithm.

Step 2. Let $i = 1$.

Step 3. Delete the connection between a_i and b_i and then apply the maze running algorithm to points a_i and b_i.

Step 4. If $i = n$ the algorithm completes. Otherwise, let $i = i + 1$ and go to step 3.

6.3. Ordering Nonplanar Connections to be Constructed

For simplicity consider the problem of selecting the optimum order of nets in constructing nonplanar connections for such circuits where the net can be divided just into two sections in the planar part. Let $\{H_1, ..., H_k\}$ be the set of all nets each of which is divided into two sections. Constructing an optimal nonplanar connection between the sections of net H_1 is the problem of finding the shortest chain of weight d_1 in the weighted graph $N(R, X)$. As a result the graph $N(R_1, X_1)$ turns out to be a model for constructing a nonplanar connection between the sections of H_2. In this graph we find the shortest chain of weight d_2, etc. As a result the sequence of weighted graphs

$$N(R,X) = N(R_0, X_0) \to N(R_1, X_1) \to \ldots \to N(R_k, X_k)$$

is obtained.

Suppose nonplanar connections between the sections of these nets are constructed in the sequence of their numbers. In this case all nonplanar connections of the circuit whose 'cost' is equal to the sum $\sum_{i=1}^{k} d_i$ of weights d_i of the shortest chains found in the appropriate graphs $N(R_{i-1}, X_{i-1})$ are constructed. Therefore an 'unsuccessful' construction of some nonplanar connections may clearly lead to the fact that the 'cost' of constructing other connections will greatly increase. Hence a problem of selecting the order of nonplanar connection construction arises. It consists in numbering the nets such that the sum $\sum_{i=1}^{k} d_i$ be minimum.

Denote by d_i^0 the length of the shortest chain in graph $N(R,X)$ which connects two faces of R containing the vertices of different sections of net H_i, i.e. the minimum 'cost' of constructing this nonplanar connection regardless of the remaining connections. Let us consider a subproblem of the stated problem: is there any construction of nonplanar connections whose 'cost' coincides with $\sum_{i=1}^{k} d_i^0$? This problem is reduced to constructing K non-intersecting chains in the graph $N(R,X)$. Since a graph treated as $N(R,X)$ can be constructed on any planar graph, a well known NP-complete problem of non-intersecting chain in a planar graph [30] is reduced to the stated problem.

Proposition 6.4 *If nets to be connected can be divided in two sections in a planar part of the circuit then the problem of ordering nonplanar connections to be constructed is NP-hard.*

Below we discuss three approximation algorithms for solving this problem. In what follows we assume that each net $H_1, \ldots, H_k$ can be divided into two unconnected sections. Generalization to an arbitrary case in selecting a net can be performed in two directions:

- determine a pair of sections with the minimum shortest distance;
- determine all nonplanar connections between the sections of the selected net.

Consider the first of the proposed algorithms.

Step 1. Let $i = 1$, $N(R_0, X_0) = N(R, X)$.

Step 2. For all $j = 1, ..., k-i+1$ determine the weight d_{ij} of the shortest chain connecting the sections of net H_j in the weighted graph $N(R_{i-1}, X_{i-1})$. Let $d_{j_0} = \min\{d_j : j = 1, ..., k-i+1\}$.

Step 3. Change the numbers of nets H_{j_0} and H_j.

Step 4. Construct nonplanar connections between sections of the net H_1. Construct the graph $N(R_i, X_i)$.

Step 5. If $i = k+1$ the algorithm completes. Otherwise, let $i = i+1$ and go to step 2.

Estimate the time complexity of the algorithm. Steps 2–5 are repeated k times. At step 2 there are $k - i + 1$ determinations of the shortest net performed in a weighted graph. Since the graph's vertices correspond to the faces of a planar embedding its running time does not exceed $O(f^2)$ where f is the number of faces in the embedding R. Hence the running time is $O(k^2 f^2)$ for the algorithm.

The other two algorithms differ from that discussed above only in step 2.

In one of them, net selection is performed by the formula $d_{j_0} = \max\{d_j : j = 1, ..., k-i+1\}$ at step 2.

The running time for the algorithm is then $O(k^2 f^2)$.

In the second algorithm, a net is selected as a result of the below analysis.

Step 1. Let $j = 1$.

Step 2. Find the shortest chain in the graph $N(R_{i-1}, X_{i-1})$ which connects the sections of net H_j. The weight of this chain we denote by d_j. Construct a nonplanar connection corresponding to the chain and go to graph $N(R_{i-1}, X_{i-1})$.

Step 3. For each $s = 1, ..., k-i+1, s \neq j$, find the weight d_{k_j} of the shortest chain in a weighted graph $N(R'_{i-1}, X'_{i-1})$ which connects the sections of net H_k. Let $D_j = \sum_{k \neq j} d_{kj} + d_j$.

Step 4. Delete nonplanar connections of H_j sections and go to the source graph $N(R_{i-1}, X_{i-1})$.

Step 5. If $j = k - i + 1$, let $j = j + 1$ and go to step 2. Otherwise find

$$D_{j_0} = \min\{D_j : j = 1, ..., k - i + 1\}$$

and the algorithm completes.

The time complexity of the algorithm for ordering nonplanar connection construction is $O(k^3 f^2)$.

Chapter 7

Planarization System Structure

7.1. Purpose and Functions of the System

The implementation of the research described in previous chapters is discussed in the present chapter devoted to a real system for synthesizing a plane embedding of a circuit which makes part of a VLSI CAD system. According to the purpose and functions of the system, possible solutions of the main problems of system design are suggested: the contents and interaction of the system program base and data organization.

Enquiries into the planarization problem that have been discussed in Chapters 2 to 6 show that its implementation in practice consists, on the one hand, in the development of an adequate mathematical model. On the other hand, the circuit planarization problem can be partitioned into independent subproblems which are mostly intractable. Depending on the class of VLSI to be designed one can select not just graphs or a special type of hypergraphs as a model, but far more complex element models discussed in Chapter 4. For lack of an efficient and precise algorithm for VLSI planarization, the selection and development of approximation algorithms are currently the urgent problems.

Therefore, when selecting a mathematical model for a VLSI circuit in the planarization problem we have to find a natural tradeoff between relatively simple (e.g. graph) and complex (e.g. hierarchical) models. Simple models allow us to apply analysis and processing methods which are efficient from the computational point of view but take no consideration of some important VLSI properties, for example the assignment of physically and logically equivalent pins, which cannot but influence the resultant efficiency. Complex universal models, even if adequate enough for planarizing a wide range of VLSI, require many computational resources for testing and processing, thereby causing performance degradation, and hence force us to apply more simple approximation methods for solving intractable problems arising in VLSI planarization, which in its turn affects the planarization result.

Thus the resultant efficiency can be lost either as a result of incomplete of testing caused by the 'simple' model inadequacy or by the forced application of 'simple' approximation algorithms for analysing 'complex' models since more complex approximation algorithms cannot be applied for such models because of a sharp increase in computation requirements.

The development of a VLSI planarization system which can be automatically tailored to wide classes of models and algorithms will allow us, in particular, to eliminate the necessity of writing many specific programs for planarizing different VLSI types. It is evident that a truly universal system which allows us naturally to incorporate *a priori* random planarization algorithms and analyse any models can not be developed, but there exists a possibility of covering a wide range of both iterative algorithms and models.

We shall first find out these reasons.

When analysing the iterative algorithms for VLSI planarization in Chapter 5 we stated that alternative solutions arise at the initial step of source cycle selection in a graph or hypergraph model, at each iteration step for selecting a successive embeddable segment and external path, as well as at min-cut routing. As has been demonstrated in [17], the result obtained by the iterative algorithm does not depend on alternative solutions in graph planarity testing. In contrast, the approximate solution of the planarization problem both for hypergraphs and graphs depends substantially on the selected alternative. It also turns out that the choice of an optimum alternative solution normally leads to an NP-complete problem. By varying the approximation algorithms for solving the above subproblems we obtain a wide class of approximation algorithms for extracting the maximum planar part of the circuit. The algorithms thus derived have common iteration scheme.

With regard to models, we may say that the hypergraph connection model is universal. The variety of VLSI circuit models arises when a hypergraph connection model is merged with the element models. In this case element models may be analysed locally by deciding whether the order of connections routed to an element is valid, whether over-the-element routes are possible, etc. The general iterative planarization scheme is actually applied to a hypergraph or graph connection model, and the feasibility of any partial embedding is checked by local element model analysis.

Actually, the above discussion concerns the first stage of circuit planarization, which involves either the embedding of the whole circuit model (i.e. circuit planarity is tested) or some of its maximal part. However, the results of Chapter 6 show that a similar situation exists at the second stage when nonplanar connections are implemented by over-the-element routing and connector placement. Depending on the technology, diffused tunnels, jumpers, vias, etc can be used as connectors. Since the connector layout may strongly depend on process conditions the system will presumably comprise different layout computation programs.

The above may be summed up as follows. The circuit planarization system is designed to:

- test a circuit for planarity;
- extract and embed in the plane its maximum planar part;
- implement nonplanar connections by over-the-element routes (for bipolar VLSI diffused tunnels serve as connectors).

Concurrently it solves (in an approximate way) one of the folowing optimization problems:

- minimization of connectors or their total length with the specified constraint on the length of a single element;
- minimization of the maximum connector length with the specified constraint on the number of elements;
- minimization of over-the-element routes.

System invariance relative to the initial mathematical model variant allows us to take account of the following process and metric constraints for different VLSI classes:

- the order of placing pins on the boundary of the element and the sequence of external pin areas;
- physical (duplicate) and logical equivalence of pins in the elements and external pin areas;
- permanent orientation of some elements;
- metric relations in an element during over-the-element routing.

The possibility of generating various iterative algorithms for VLSI planarization allows that a most efficient approximation algorithm is chosen for specific class of circuits.

The planar circuit embedding produced by the system is used as an initial sketch for VLSI layout design. At subsequent stages, the design can be manual, semi-automatic or fully automatic.

7.2. Planarization System Structure. Logic Diagram

Definition of nets and circuit element models serves as input to the planarization system. In the first stage, the mathematical model for a circuit is produced from data in the integrated database of the automated VLSI design system.

As mentioned earlier, the generated mathematical model is the 'union' of the hypergraph connection model with the element models. To reduce the size of the circuit planarization problem to be solved, various decomposition procedures, from the simplest one such as deletion of articulation vertices and bridges, to fairly complex such as those discussed in Chapters 2 and 3, are used in the second stage. These procedures reduce the initial circuit model to the planarization of some collection of 'irreducible' models which will be further applied to 'immediate' planarization methods.

In the third stage the system extracts and embeds in the plane the maximal planar part of a VLSI circuit. For this purpose various iterative algorithms for hypergraph planarization are applied (refer to Chapter 5). In view of this stage's importance we shall concentrate on the estimation of its running time and storage requirement.

The top asymptotic running time estimates for iterative hypergraph planarization algorithms which strongly depend on the approaches to solving subproblems (choice of an initial cycle, an embeddable segment, etc.) were discussed in Chapter 5. The statistical averages of planarization of real circuits have shown that the time for extracting the maximal planar part is practically proportional to the square of a number of pins involved in the elements. In such cases the algorithms with a quadratic top estimate of run time were chosen for solving the subproblems. Hierarchical models (see Chapter 4) were used as element models.

To extract and embed the maximal planar part of a circuit in the plane, the element models, lists of hyperedges associated with the circuit nets are loaded into memory. At each step of iterative algorithm execution there is a request for storage to generate the segment list. The number of segments obviously does not exceed the number of edges in the hypergraph model. The number of faces in the planar embedding, the number of elements, nets and logical pins do not exceed the number of physical pins in the circuit. Thus the main storage size required for extracting the maximal planar part of the circuit linearly depends on the number of physical pins in the circuit.

To optimize the extracted maximal planar part, local operations can be applied. Local optimization of circuit embedding acts here the same as the methods used to improve the initial type placement of the pairwise permutation algorithm when solving the classical element placement problem. We may distinguish between three kinds of local operations (item 5.4):

- net reconfiguration;
- biconnected component pivoting around articulation point and moving 2-connected and 3-connected components to another face;
- substitution of some connections with a greater number of earlier nonplanar connections.

Local optimization operations are performed iteratively on an embedding until the embedding can be improved by at least one of them. The order of local operations is chosen as the basis of the 'greedy' algorithm, i.e. the operation selected to be executed must produce the maximum effect.

At the final stage of planarization nonplanar connections are constructed by over-the-element routing and connector placement. The mathematical model for such a problem is a weighted graph of nonplanar connections (refer to Chapter 6). Nonplanar connections may be constructed by the methods discussed in Chapter 6. The procedure can be divided into two basic stages: selection of the order of nonplanar connection routing and routing a nonplanar connection. Statistically average running time for these algorithms turns out to be proportional to the square of the sum of the number of enabled element pins involved, the number of routes over the connectors, and the number of over-the-element routes. Memory size required for the purpose as well as that required for extracting the maximal planar part linearly depend on the above sum.

When 'irreducible' circuits obtained by partitioning an original one have been planarized a plane embedding of the original circuit is generated from the obtained plane embeddings of 'irreducible' circuits.

Figure 7.1 shows the planarization system scheme.

The program base for the VLSI circuit planarization system thus consists of five program systems (Figure 7.2).

1. Generation of a circuit model which adequately fits its planarization problem.
2. Circuit decomposition. The programs which make part of this system 'split' an original circuit into 'irreducible' subcircuits and generate the plane embedding of the original circuit from the plane embeddings of 'irreducible' subcircuits.
3. Extraction and embedding the maximal planar part of the circuit in the plane.

   The core of the system form the programs which implement the iterative

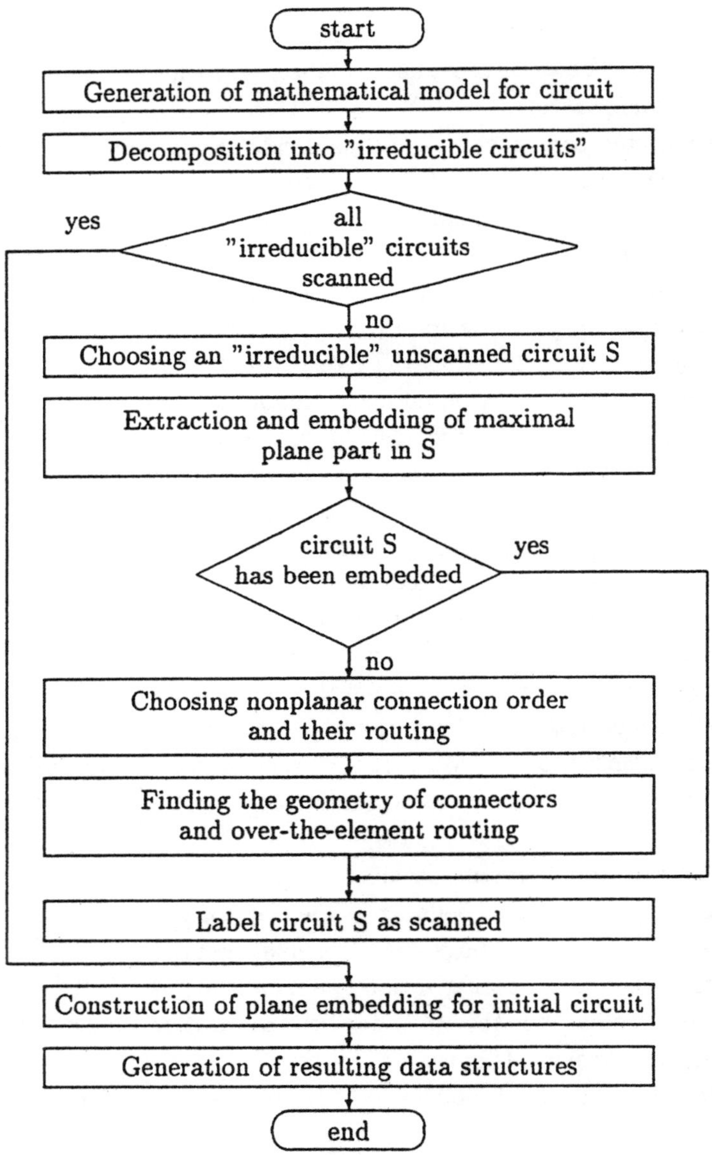

Fig. 7.1. Planarization system logic diagram.

algorithm for planarizing the hypergraph model of the circuit. These are primarily the following programs:

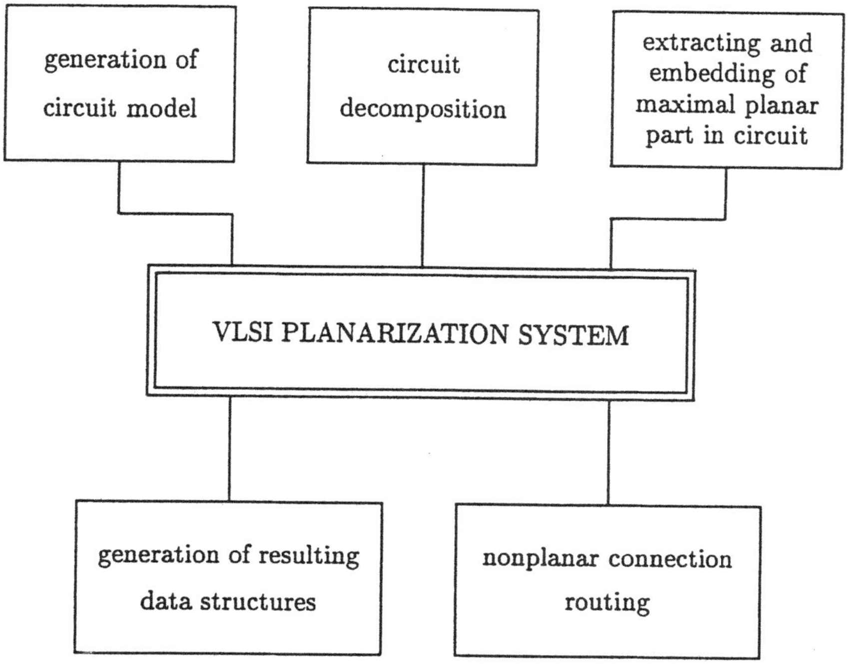

Fig. 7.2. Program base for planarization system.

- extraction and embedding of an initial cycle;
- determination of segments;
- segment embeddability testing;
- extraction and embedding of an external net in a segment;
- unembeddable segment splitting.

As a result of running the above programs we obtain the definition of a plane embedding for the circuit.

4. Constructing nonplanar connections for a circuit.
   This program system constructs nonplanar connections by over-the-element routing and placing connectors. The principal programs of the system are generation of a minimum tree for spanning the circuit sections and nonplanar connection sequencing. The system also comprises the programs for checking over-the-element routes for feasibility and optimize embedding by local operations.
5. Generation of resulting data structures.

The programs of this system are run at the final stage and form an interface between the planarization system and the integrated database of the automated VLSI design system.

Figures 7.3 – 7.11 illustrate the system logic.

7.3. Data Organization in Planarization System

Data organization in a planarization system involves the following steps:

- representation of element models;
- representation of a hypergraph model for connections;
- representation of a plane embedding.

Let us consider each of the steps in more detail.

If a graph element model is used for planarization, the model is united with the hypergraph connection model which will be discussed below. Suppose that a hierarchical model of elements is used. In this case, the element type, permissible orientation, and overall dimensions are stored for each element. We shall use the list structure to represent a hierarchical model. Each element is represented as a list of classes of logically equivalent pin groups. Each class in its turn is represented as a list of its pin groups, each group being formed by the list of logically equivalent groups of pins, etc. A logical pin is represented as a list of locations for the physical pins which form the logical pin. For each element a list of physical pins in the order of their arrangement on the element boundary is also stored. A record associated with each physical pin contains the following information:

- the maximum number of over-the-element routes between two adjacent pins;
- the cost of the route between two adjacent pins.

When an over-the-element route is required the list of physical pins is added to the records corresponding to the endpoints of over-the-element route. Each record contains:

- type,
- location of plane embedding edge incident with the pin,
- coordinates.

Figure 7.12 shows the representation of a hierarchical element model. Memory size required for its allocation is $O(n)$, where n is the number of physical pins in the element.

A hypergraph connection model is given as a Konig representation. For its storage the adjacency lists (refer to item 1.3) which contain not only the

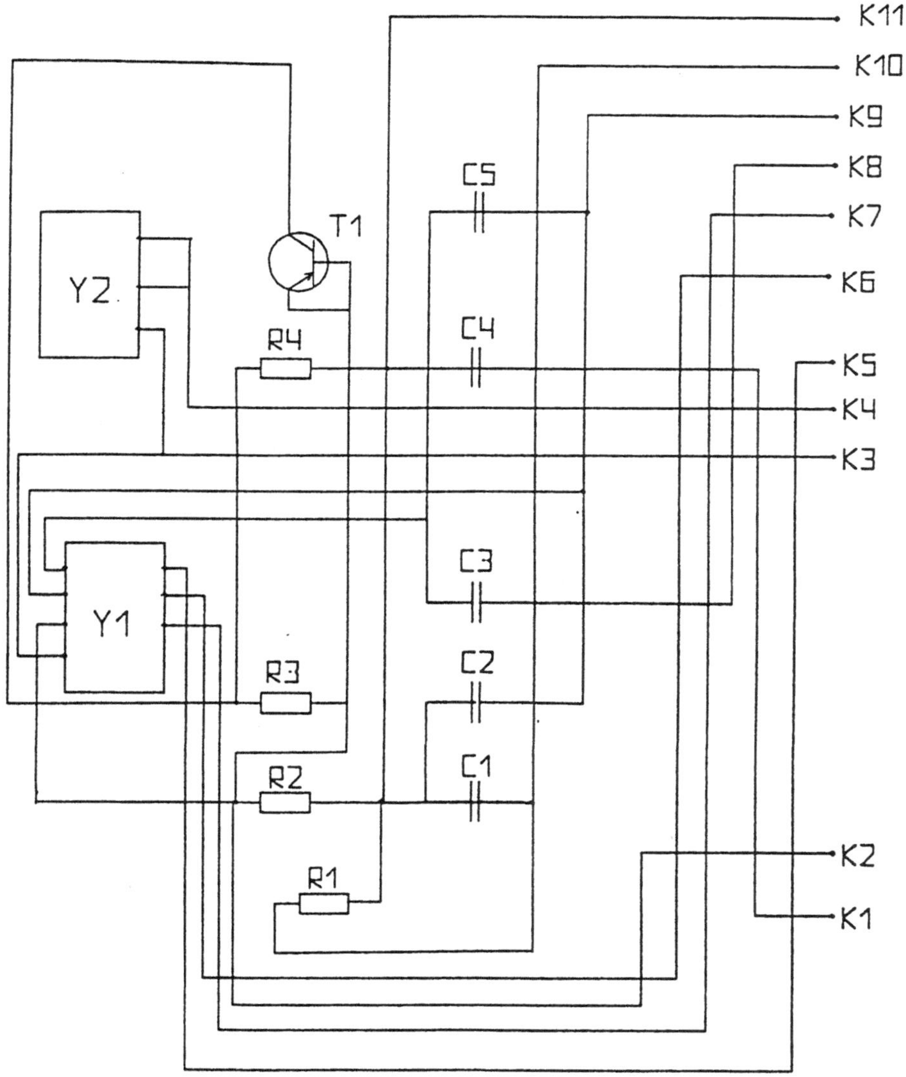

Fig. 7.3. Circuit.

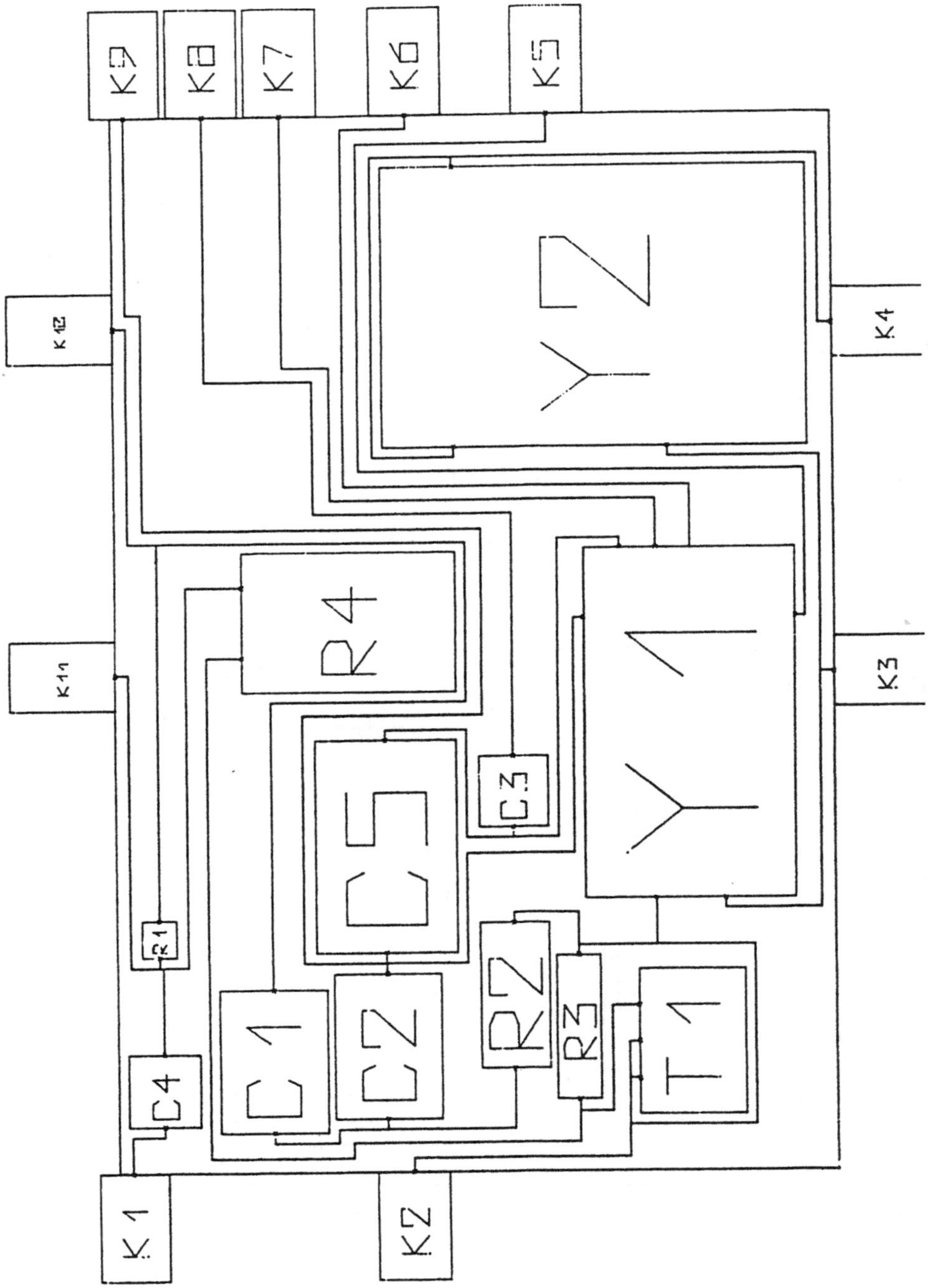

Fig. 7.4. Planar embedding of the circuit of Figure 7.3.

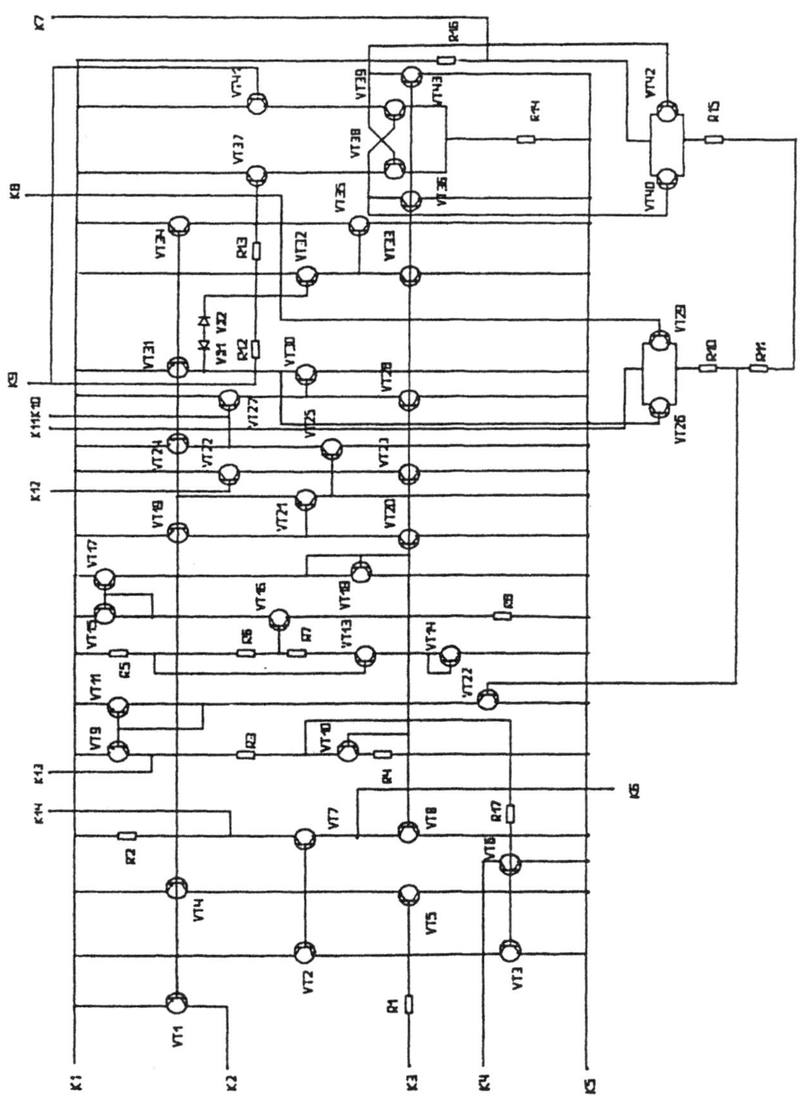

Fig. 7.5. Circuit.

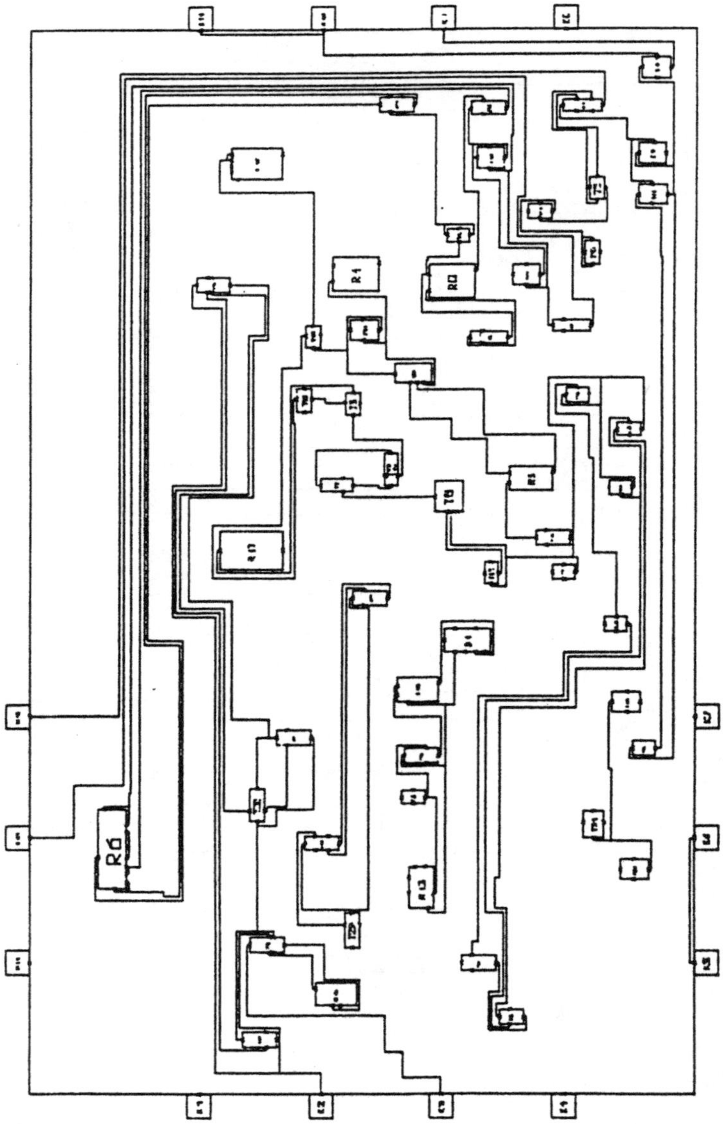

Fig. 7.6. Maximum planar part of the circuit of Figure 7.5 .

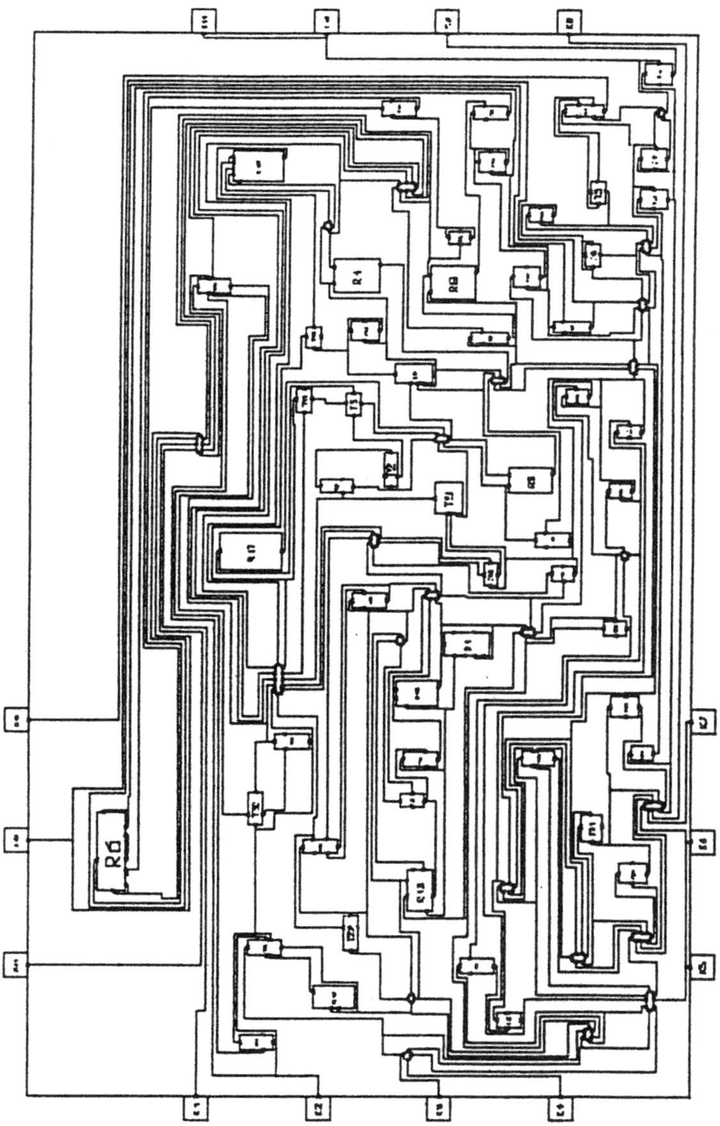

Fig. 7.7. Planar embedding of the circuit of Figure 7.5 .

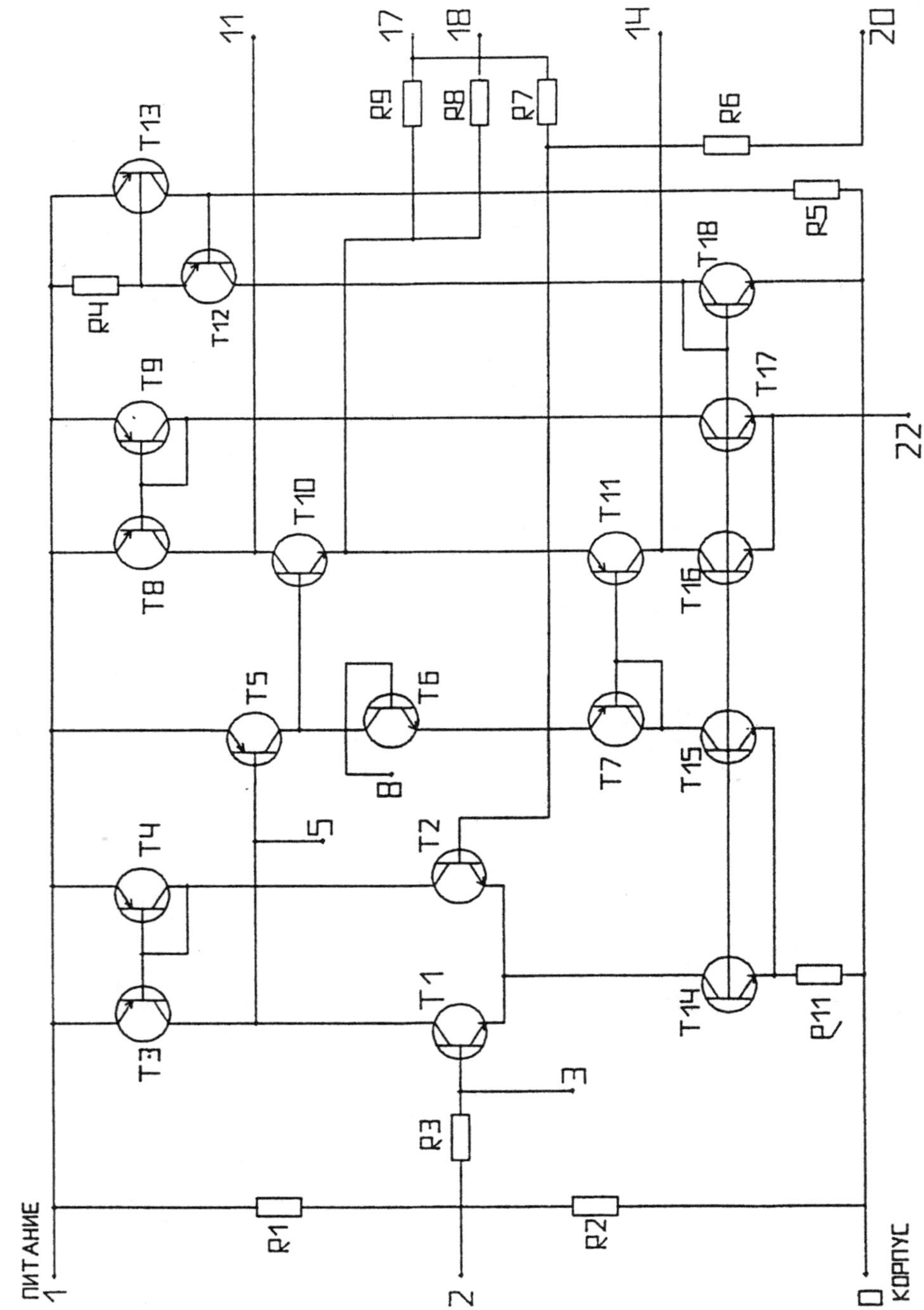

Fig. 7.8. Circuit.

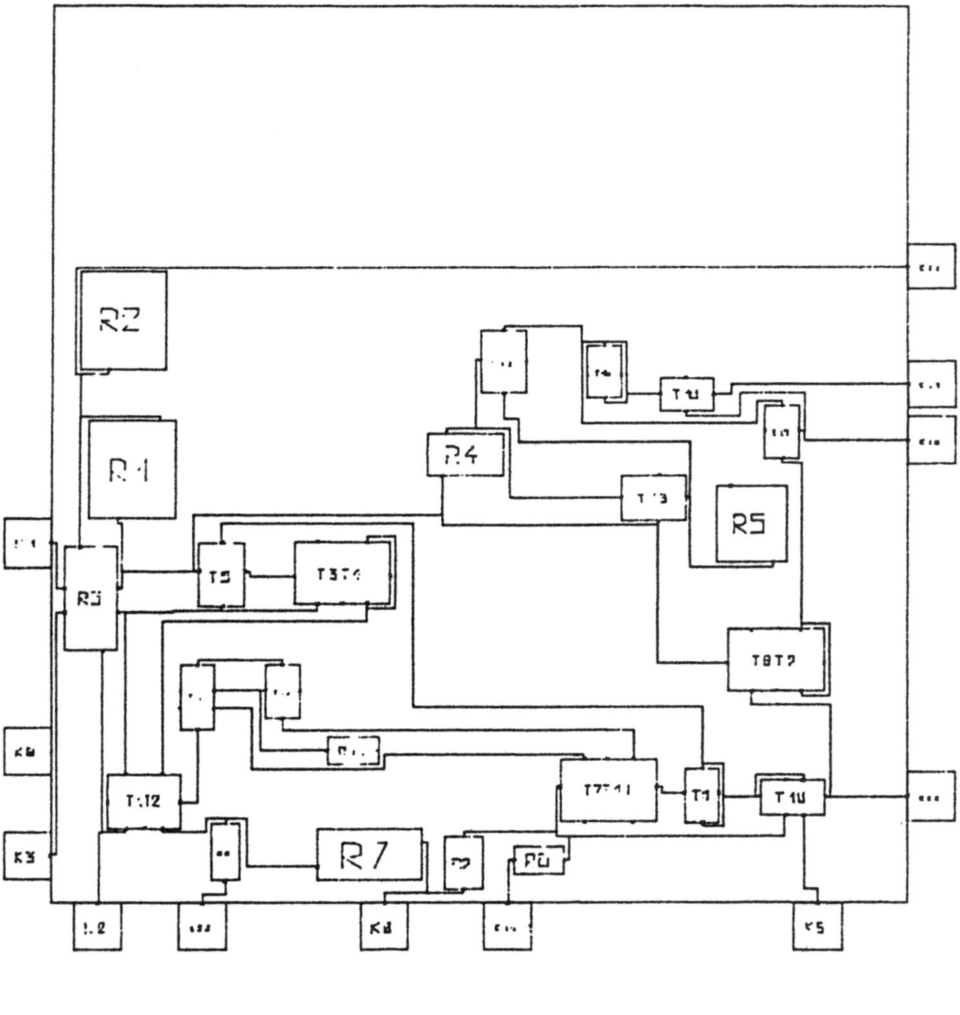

Fig. 7.9. Maximum planar part of the circuit of Figure 7.8 .

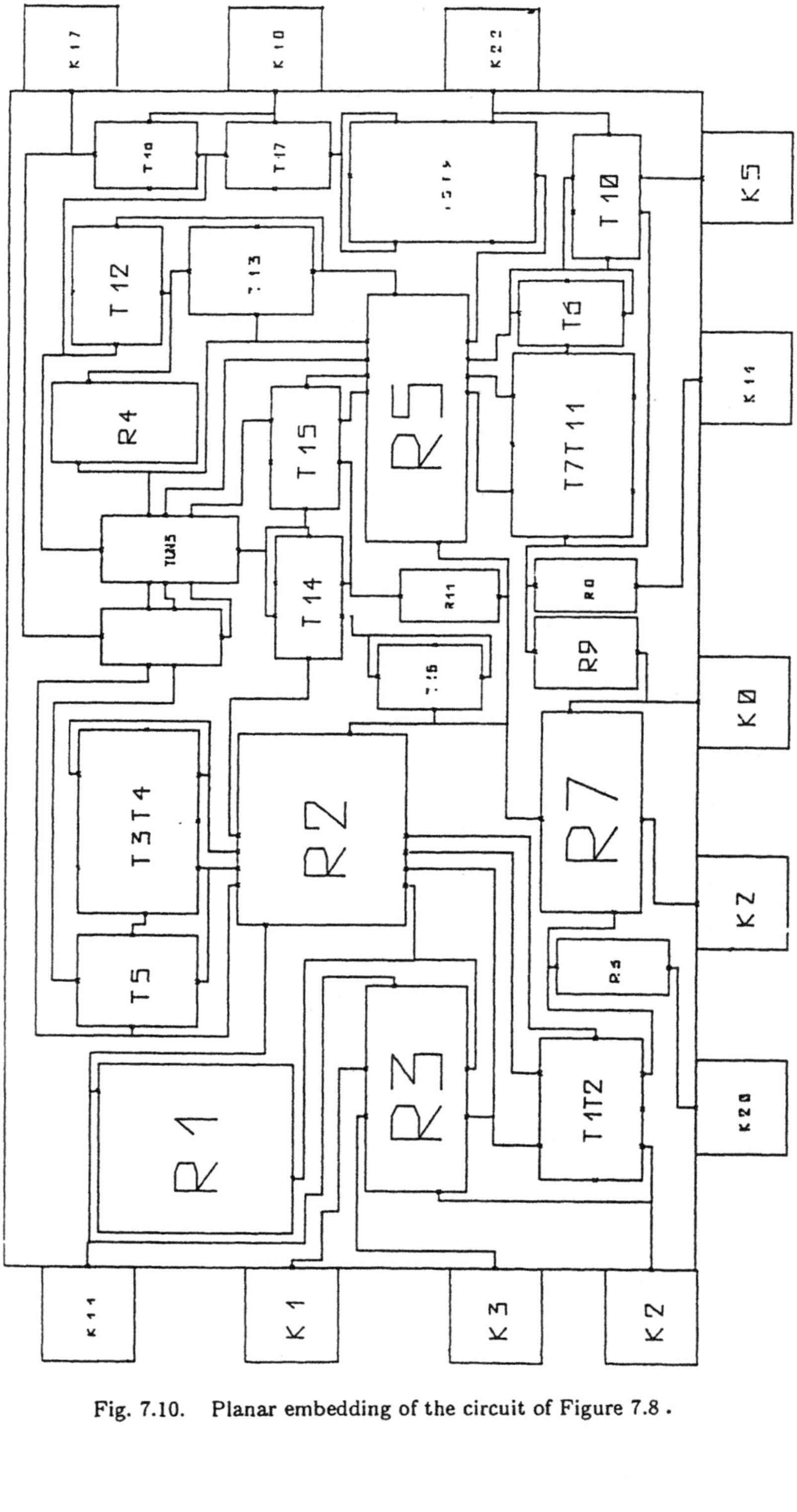

Fig. 7.10. Planar embedding of the circuit of Figure 7.8 .

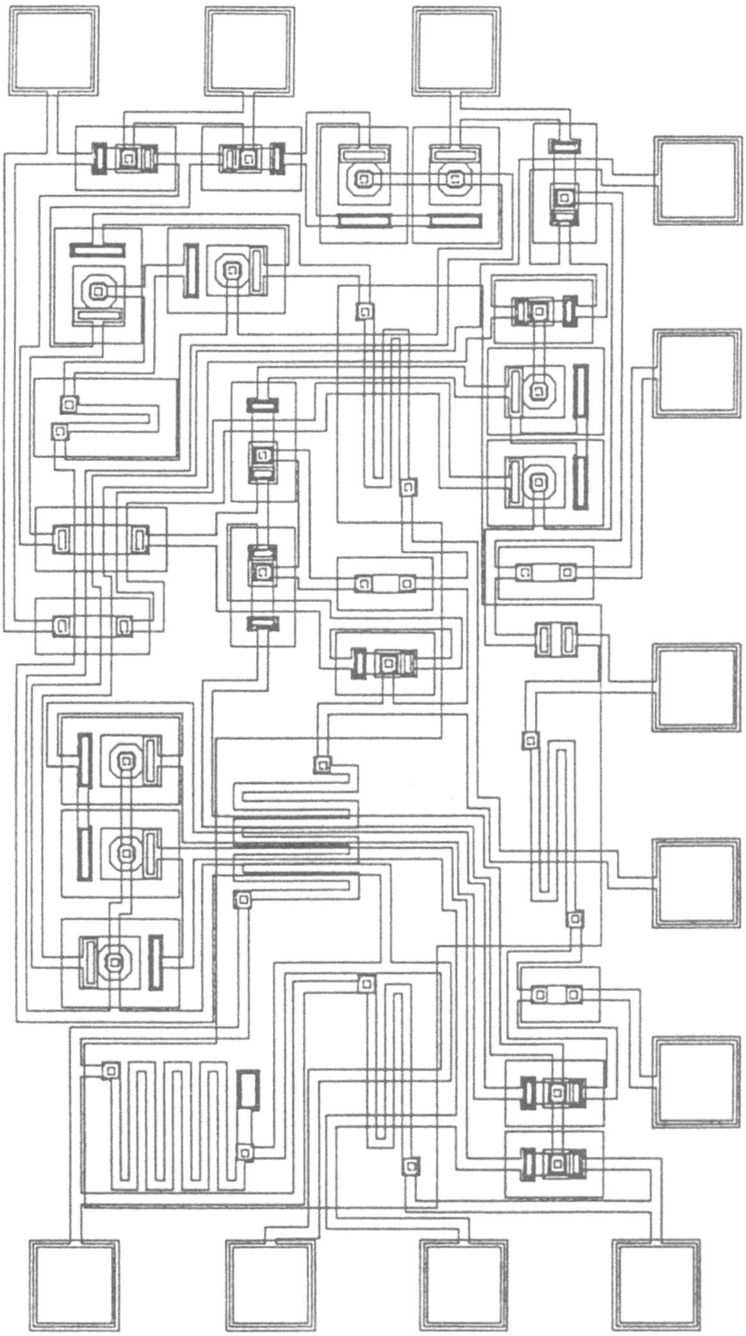

Fig. 7.11. Layout of the circuit of Figure 7.8 .

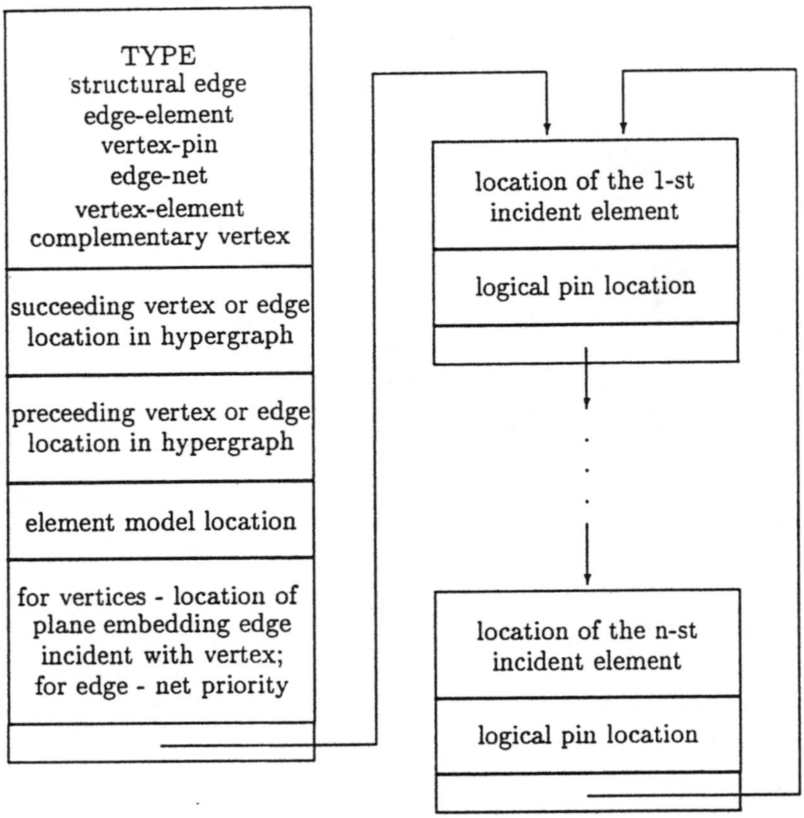

Fig. 7.12. Hypergraph connection model scheme.

location of the Konig representation vertex adjacent to the given one, but also the location of an appropriate pin of a circuit element (for the case of a hierarchical model) are used. For each vertex the memory also contains its type (vertex-element, vertex-pin, complementary vertex, etc.) and the location of a hierarchical model for the appropriate element (if necessary). For the edges of the hypergraph model, the memory contains the type (edge-net, edge-element, structural edge, etc.). Thus the memory size required for allocating a hypergraph connection model is $O(N)$, where N is the total number of pins in a circuit element. Representation of a hypergraph model is given in Figures 7.13, 7.14.

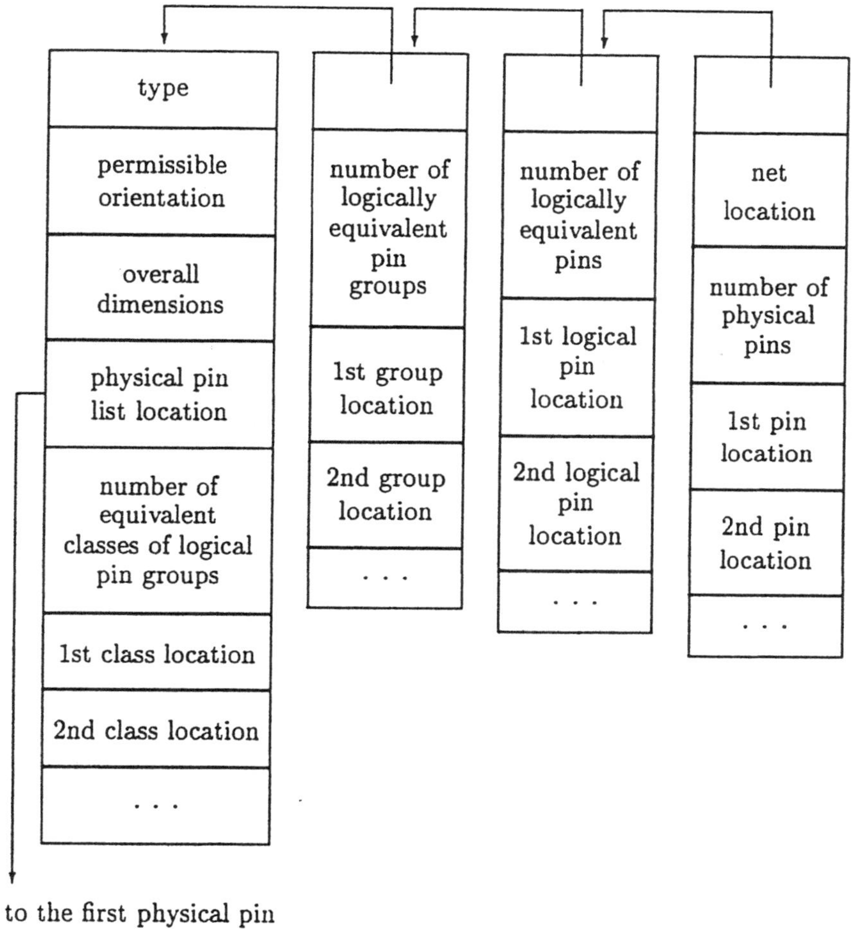

Fig. 7.13. Hierarchical model of an element (Part 1 of 2).

To represent a plane embedding we shall use the 'truncated' version of a Baumgart's 'winged' model. In this case the plane embedding is represented as an edge list and a face list. For each edge of the embedding the following data is stored (Figure 7.15):

- location of an element incident with the edge origin;
- location of the element pin incident with the edge origin (for the case of hierarchical model);

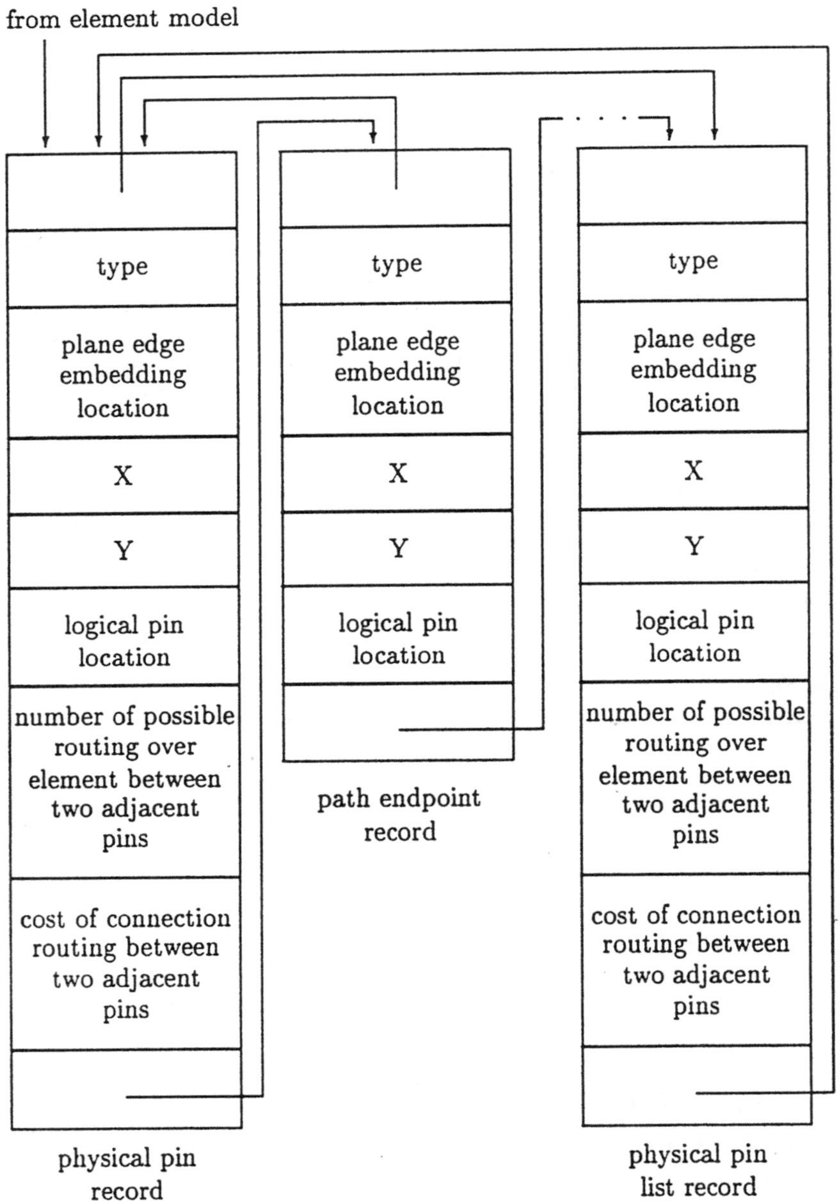

Fig. 7.14. Hierarchical model of an element (Part 2 of 2).

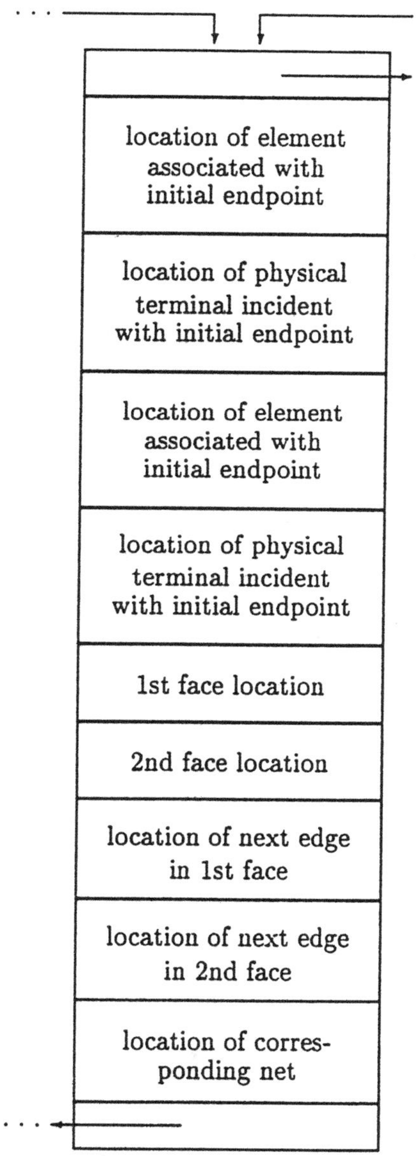

a)

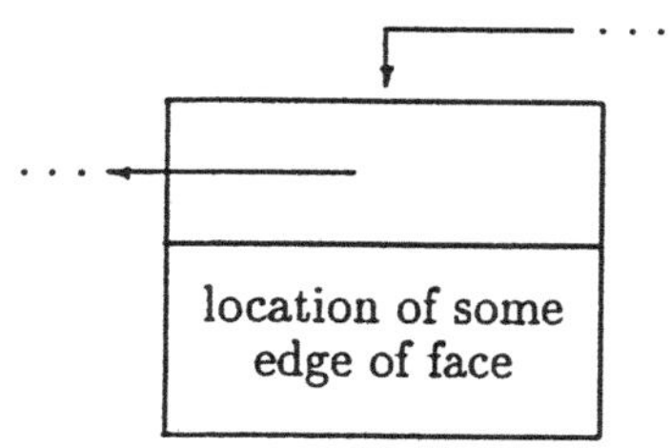

b)

Fig. 7.15. Plane embedding representation: (a) representation of edges in plane embedding; (b) representation of faces in plane embedding.

- location of an element incident with the edge end;
- location of the element pin incident with the edge end (for the case of hierarchical model);
- two faces containing the edge in question;
- a sequential edge for each of the faces;
- location of the net whose implementation involves the edge in question.

For each face the storage holds the location of some edge of the face (Figure 7.15). Such representation of a plane embedding, in our opinion, is the most optimal, since many operations on the plane embedding, such as encircling a face, generation of the list of edges incident with the given vertex, embedding of some net in the specified face, etc, become possible. To store the given representation of the plane embedding the memory size required is $O(N)$, as above, where N is the total number of involved pins of circuit elements.

To make use of element models, a hypergraph connection model, and a plane embedding, we propose to apply an API which allows us to ensure relative independence of application programs from data representation and ease their development. API may be divided into three major groups:

- interface with hierarchical models of elements;
- interface with hypergraph model of connections;
- interface with plane embedding.

The suggested solutions were checked out in practical use of a VLSI CAD system implemented on a VAX platform.

7.4. Some Aspects of the Practical Use of the Planarization System

Hypergraph planarization methods presented in the book are the basis of a VLSI Layout Planarization System which is one of the major subsystems of an industrially implemented auto-interactive VLSI layout CAD system.

Development and implementation of a viable industrial VLSI layout CAD system working in conditions of high volume manufacturing basically for analog and analog-digital VLSI design is impossible, without taking into account the real design flow which is defined by the next main factors:

- existence of specific restrictions on layout which cannot be formalized in a simple way at the stage of mathematical model formation and specification development;
- absence of an acceptable pure automatic design methods (by minimum area and number of the connector criterion) for nonregular circuits with

elements of different size;

- absence of an automatic layout design methods for analog circuits;
- necessity of taking into account an effective designer's methods which can hardly be formalized at the mathematical model formation stage;
- absence of circuit models, use of which could allow us to simulate adequately the real design process;
- large volume of layout corrections as a result of specification and modification of the technological manufacturing conditions.

The factors listed are the main ones for the definition of the structure of a VLSI layout CAD system. Actually only an interactive system allows us to take into account the abovementioned conditions of a real VLSI design and manufacturing process for analog and analog–digital VLSI.

These circumstances led one of the authors, V. Feinberg, to detailed development of an auto-interactive synthesis conception for such a class of VLSI.

The auto-interactive system gives the possibility of obtaining a design quality comparable with that developed by a highly experienced designer on a purely interactive system. The design time is significally reduced by an auto-interactive system.

For layouts with a single routing layer and one and a half routing layers the conception of an auto-interactive synthesis suggests an effective solution of the following problems on the same information, linguistic and methodical base.

1. Construction of a circuit's mathematical model.
2. Planarity testing, constructing a circuit plane embedding and nonplanar connection.
3. Constructing the original layout sketch.
4. Original sketch optimization by auto-interactive procedures of the layout deformation on the basis of a wire space continuous model.

The problems 1, 2 and 3 are solved by a VLSI Layout Planarization System on the base of the theory and methods presented in this book. Let us describe in a nut shell an approach for solving problem 4 accepted in the VLSI layout CAD system mentioned.

The auto-interactive layout optimization method is based on representating the wires as a 'rubber bands' and the elements as the moving rectangulars. This idea is implemented as a set of auto-interactive local deformation procedures. The output of the Planarization System is an input for these procedures. As was indicated above, the VLSI layout sketch obtained by the Planarization System is quasi-optimal for the number of connectors, or for the total connector length, or for the number of over-the-element routes, etc. But it can be far from optimal for area criteria.

Combination of effective procedures of computational geometry with a hierarchical representation of the circuit provides the possibility of using a hierarchical approach for VLSI layout optimization by continuous deformation method.

Thus the main object of an auto-interactive layout optimization by local deformation is a cell. According to this approach a problem of auto-interactive optimization is (as a matter of fact) a problem of editing the layout by the continuous deformation method.

The main procedures of an auto-interactive deformation are reduced to procedures of continuous local element shifting and turning, and to wire rerouting procedures.

These procedures are closely connected. Thus element shifting results in the deformation of the wires in the neighbourhood. The local wire rerouting procedures allow us to replace the deformed parts of wires automatically. It provides the possibility of consequential layout optimization. These procedures automatically keep layout and netlist matching as well as basic design rule restrictions. The procedures are basic for an auto-interactive VLSI layout CAD system. On the basic of these procedures a designer can create different macros.

An auto-interactive optimization VLSI layout CAD system based on such approach can be considered as a problem oriented layout editor and continuous deformation procedures as the specialized commands of the layout editor.

The main steps of such an auto-interactive optimization of layout are shown in Figures 7.16 - 7.29. On Figure 7.16 the circuit for a VLSI cell is presented. The plane embedding of the maximal planar part of this cell is presented in Figure 7.17. Figure 7.18 shows how nonplanar connections of this circuit are implemented by over-the-element routes and by seven diffused tunnels. In Figure 7.19 the layout sketch of the maximal planar part is shown. The layout sketch of Figure 7.20 is obtained by the system in the automatic mode. Seven shaded diffused tunnels were placed.

An intermediate version of the layout of Figure 7.21 is obtained from the layout sketch of Figure 7.19 by auto-interactive routing. The connections are implemented by the over-the-element R_1 route. Figure 7.22 shows how the additional routing was completed by placement of the diffused tunnel $TUN1$. The additional routing of Figure 7.23 was implemented by placement of the diffused tunnel $TUN2$. Figure 7.23 presents results of partial layout optimization. Figure 7.24 presents the next step of the layout design. The additional route embedding was implemented by the placement of the diffused tunnel $TUN3$. Layout was partially optimized by the auto-interactive commands TURN, SHIFT, REROUTING. The layout sketch from the Figure 7.25 was obtained from that of Figure 7.24 by the SHIFT command for the elements $TUN2$, $TUN3$,$T8$, $T9$ and the TURN command for the element $R7$.

Figure 7.26 shows the fixing of the elements. The box placed inside the element is the sign for a fixed element. Fixing was done to retain the mutual placement of the elements during further actions. The layout version of Figure 7.27 is obtained from that of Figure 7.26 by the SHIFT of the element $T5$ and by the SHIFT of the turned element $TUN1$. The layout sketch on the Figure 7.28 was obtained from that of Figure 7.27 by the COMPACTION command. The final version of the layout in Figure 7.29 was obtained in the automatic mode from the sketch of Figure 7.28 and a library component.

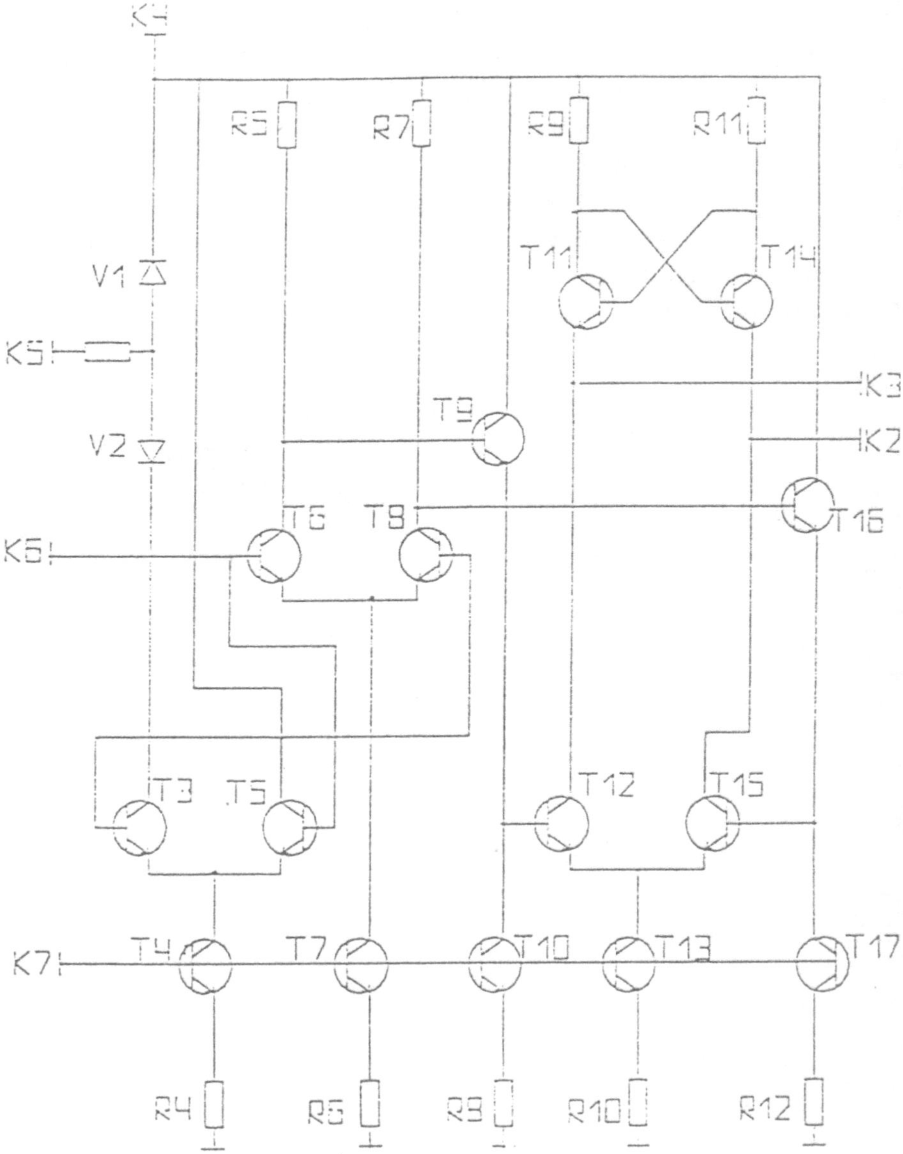

Fig. 7.16. Circuit.

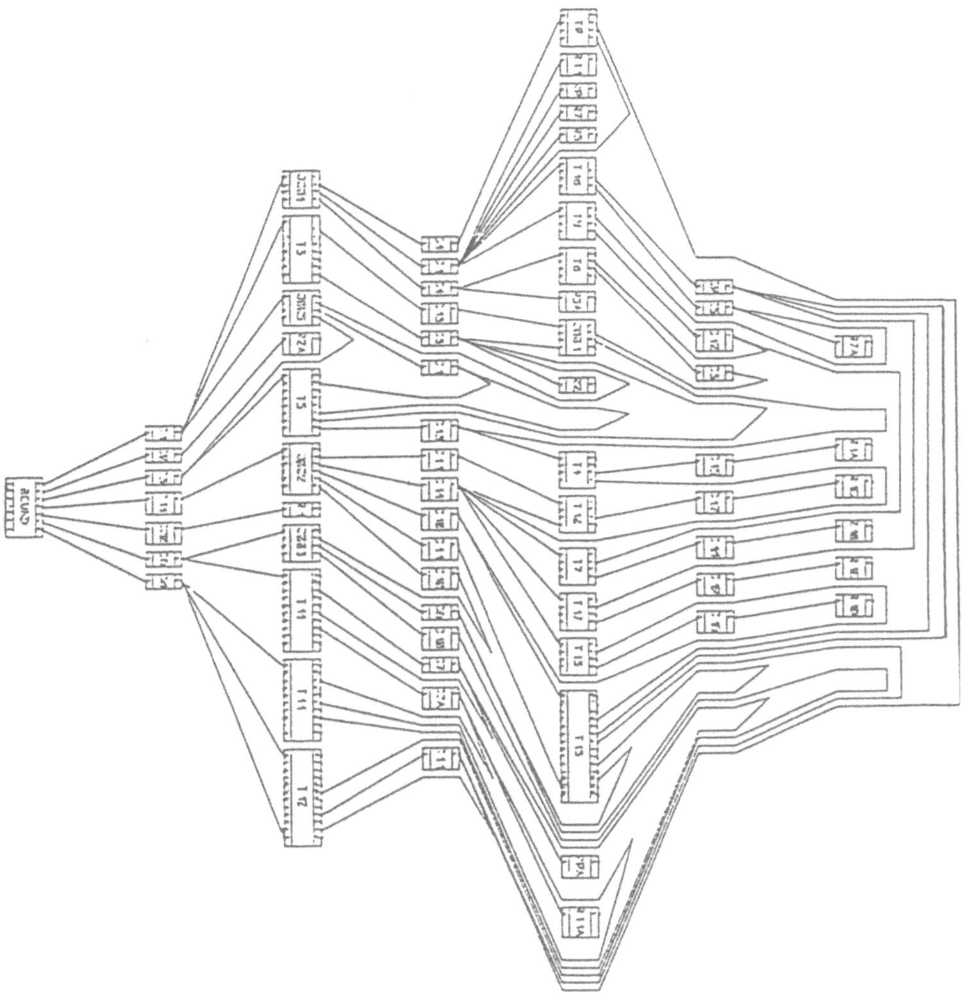

Fig. 7.17. Plane embedding of the maximal planar part of the circuit oif Figure 7.16.

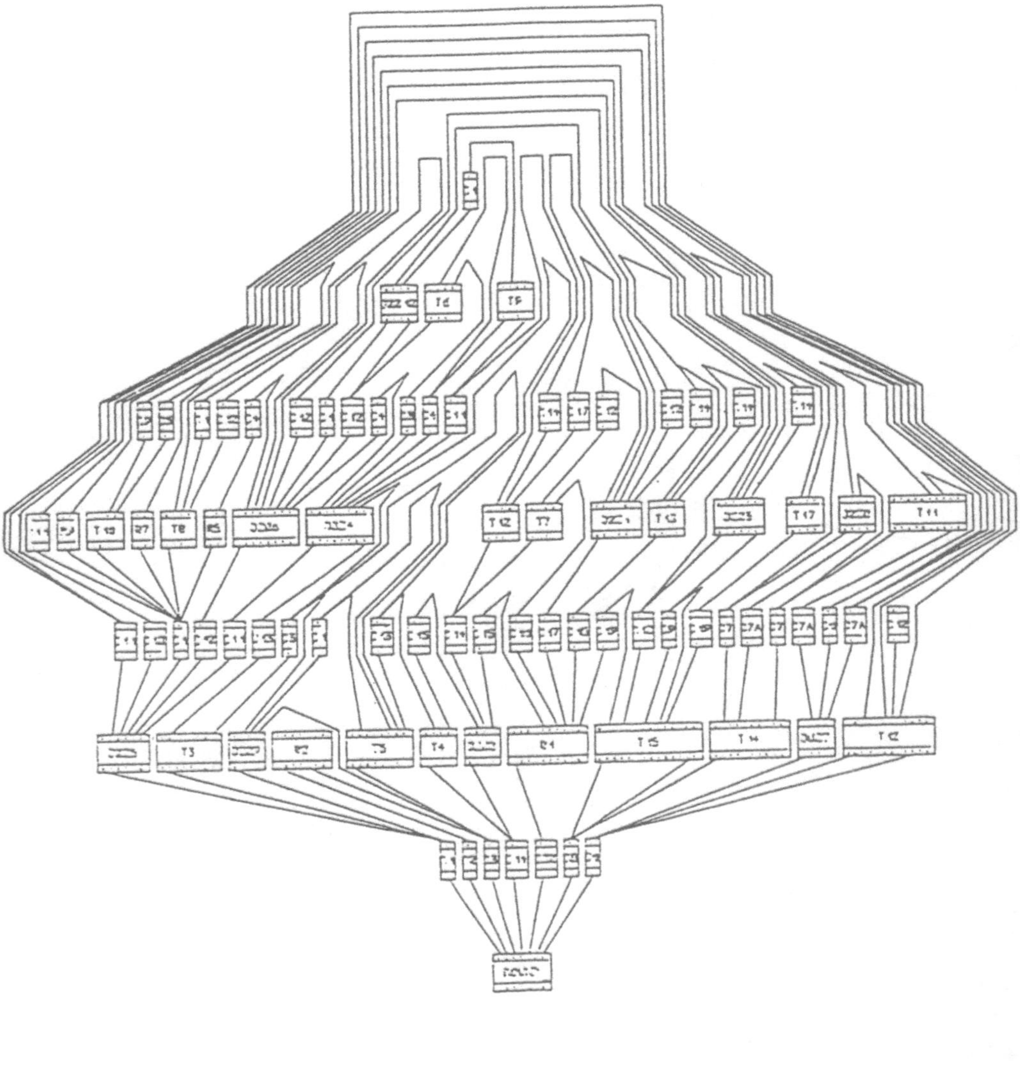

Fig. 7.18. Plane embedding. Nonplanar connections are implemented by over-the-element routes and by seven diffused tunnels.

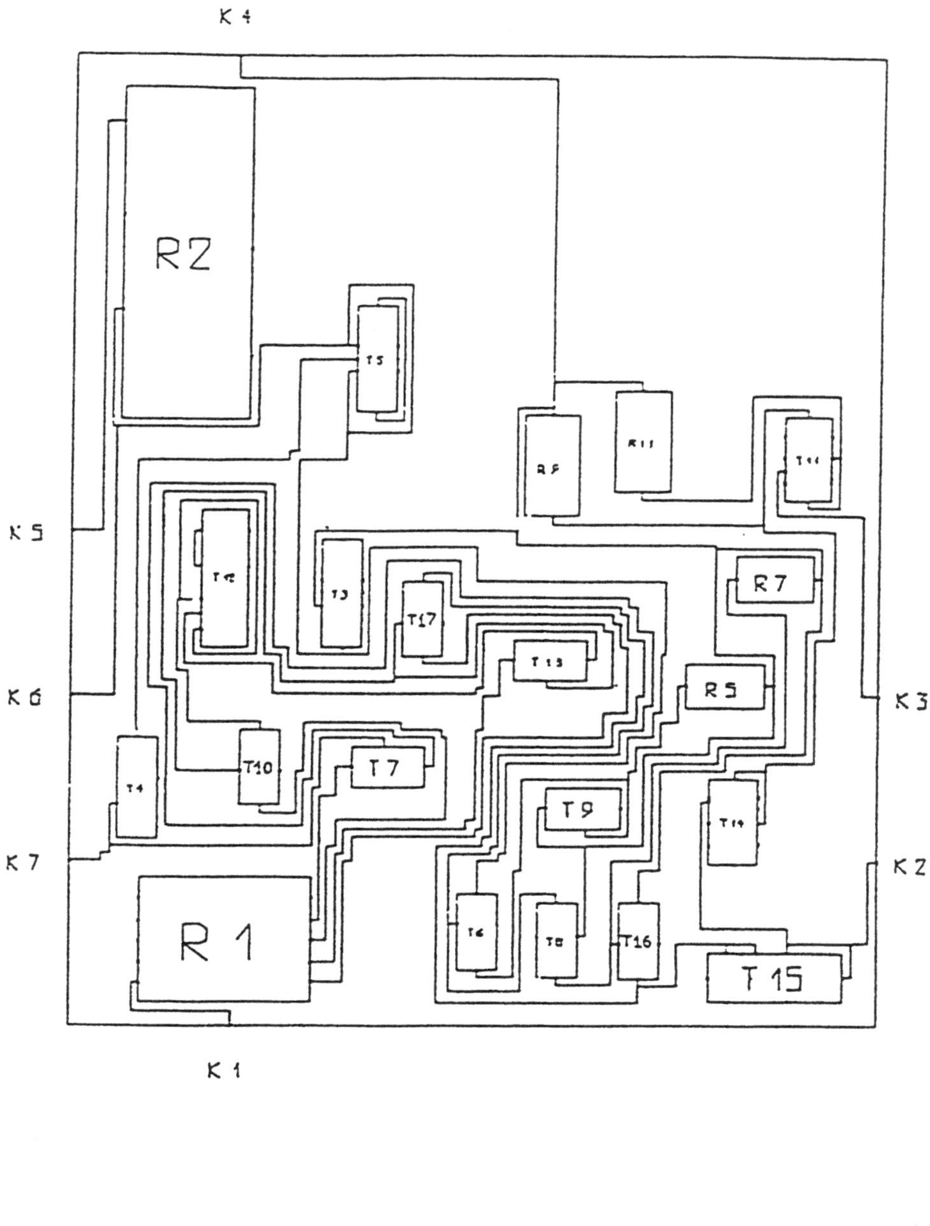

Fig. 7.19. The layout sketch of the maximal planar part.

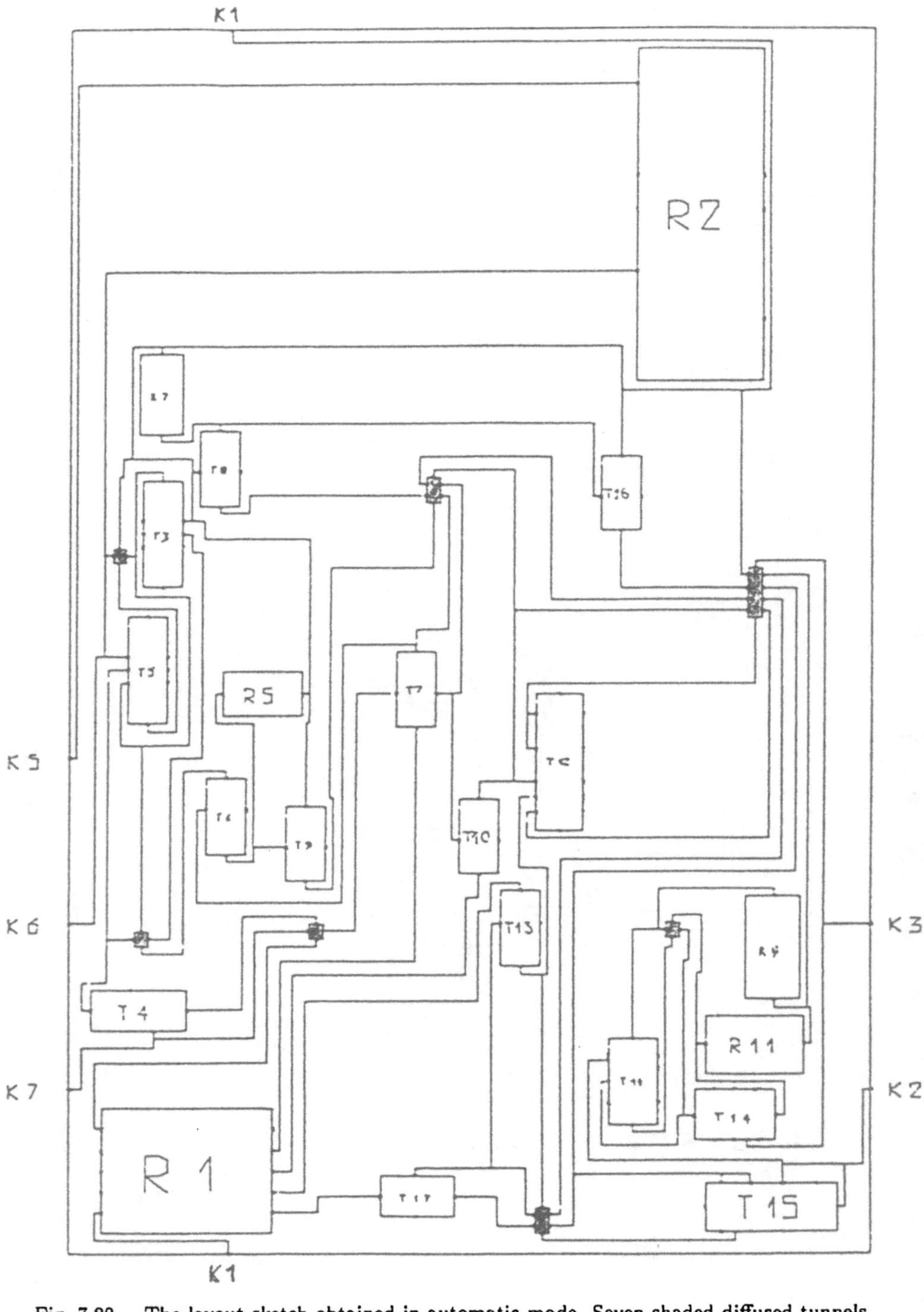

Fig. 7.20. The layout sketch obtained in automatic mode. Seven shaded diffused tunnels were placed.

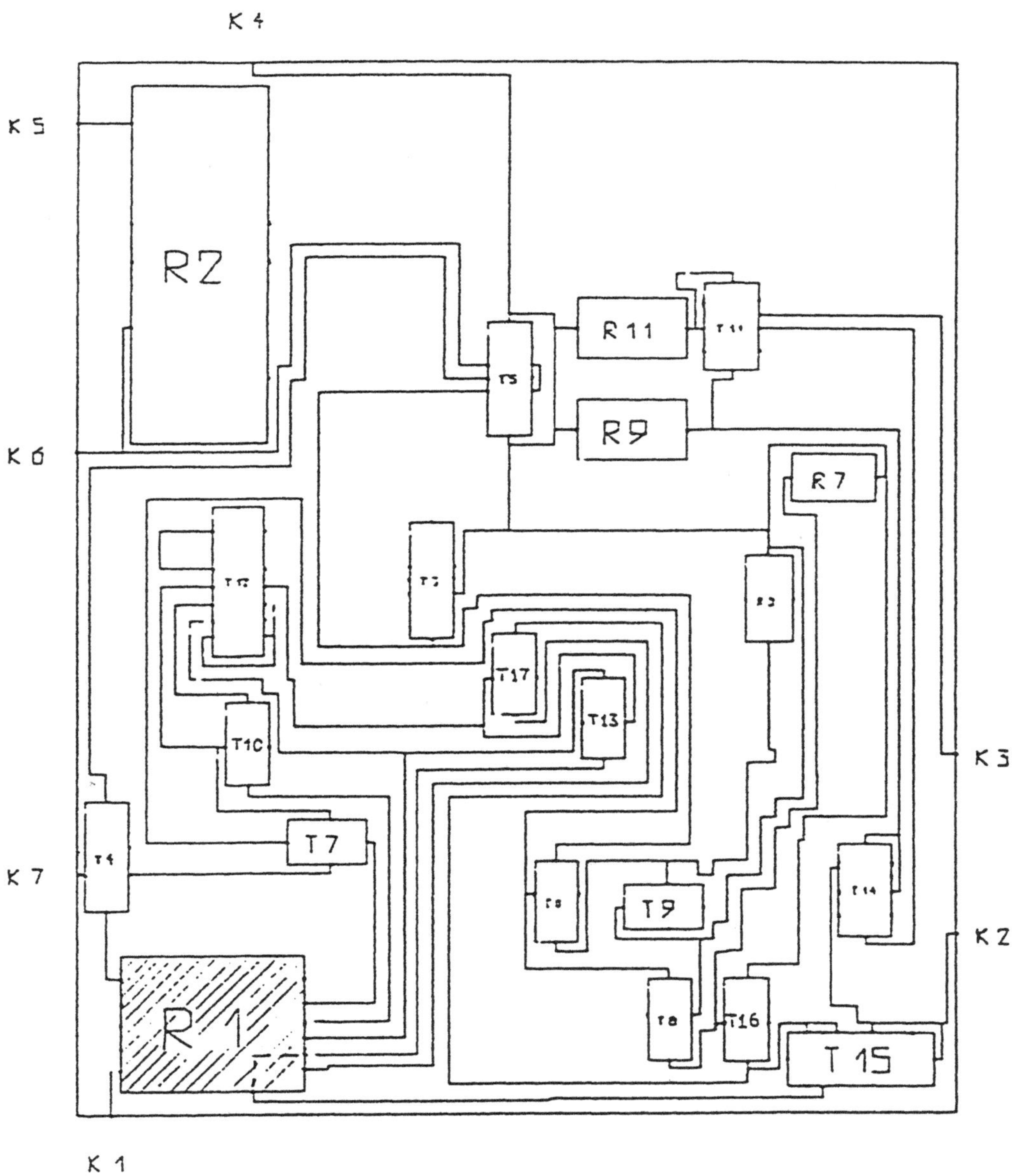

Fig. 7.21. The intermediate version of the layout obtained from that of Figure 7.19 by auto-interactive routes embedding. The connections are implemented by over-the-element R_1 route.

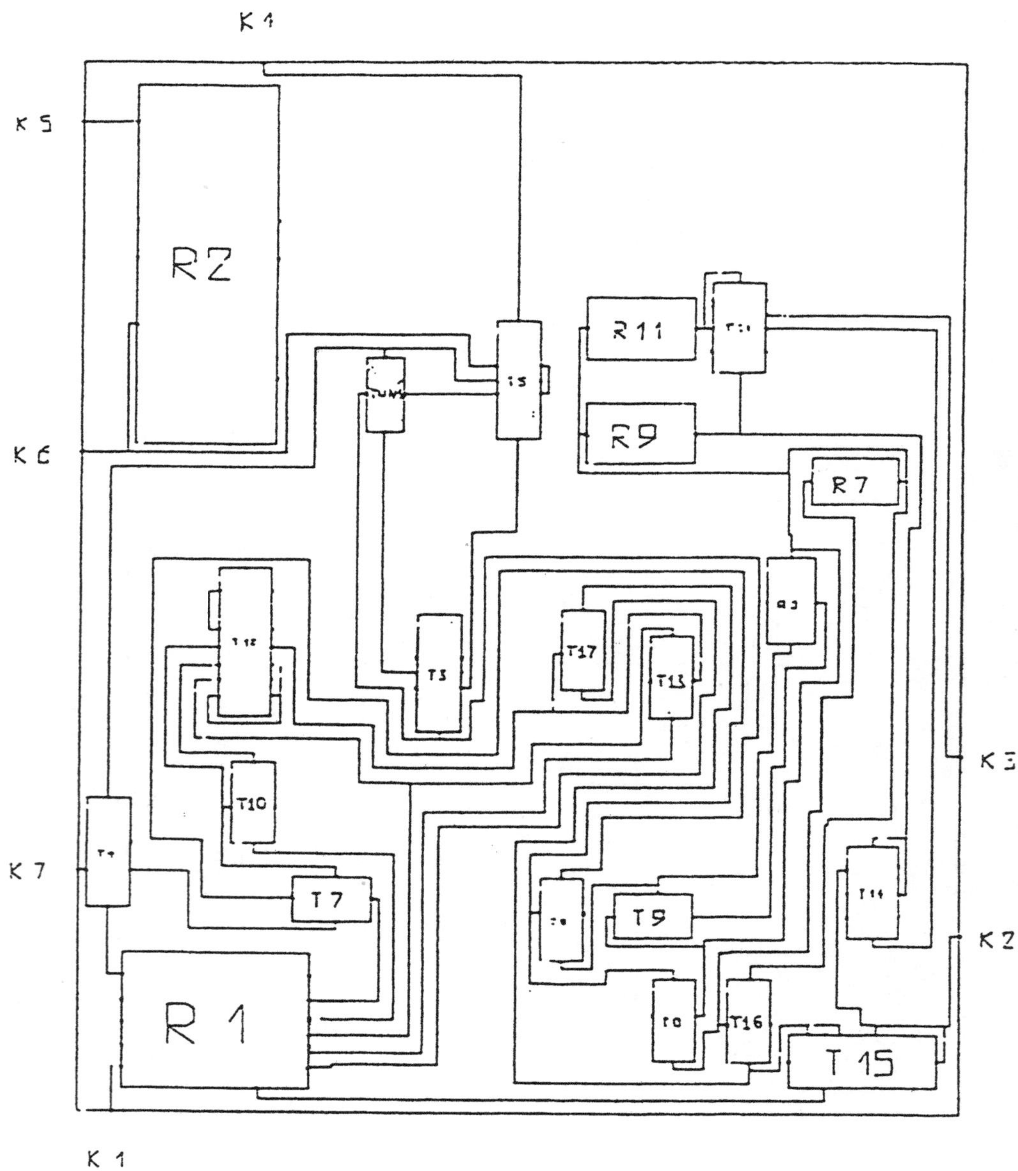

Fig. 7.22. The intermediate version of the layout obtained from that of Figure 7.21 The additional routing was completed by diffused tunnel $TUN1$ placement.

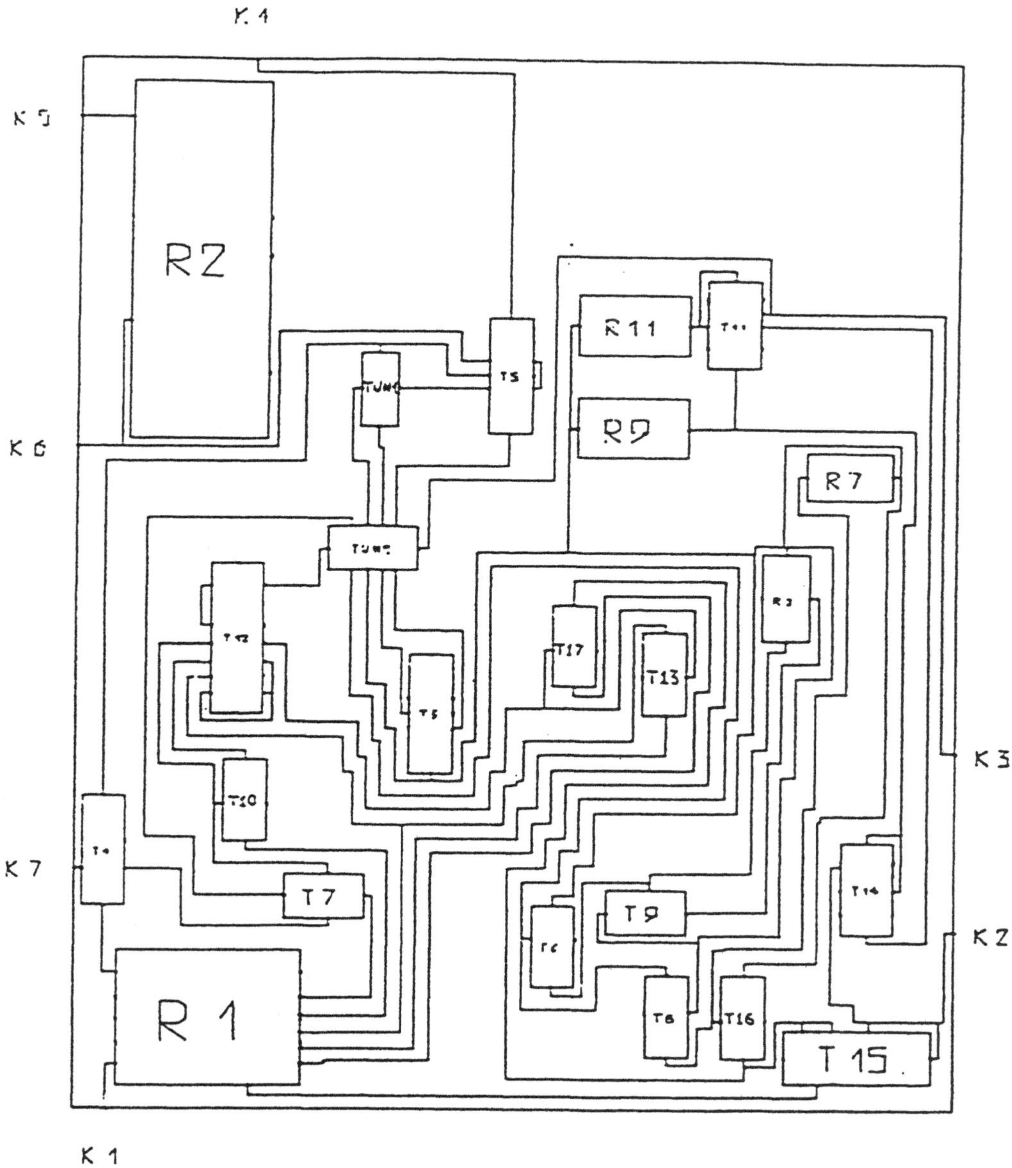

Fig. 7.23. The intermediate version of the layout obtained from that of Figure 7.22 The additional routing was implemented by diffused tunnel $TUN2$ placement. Partial layout optimization was done.

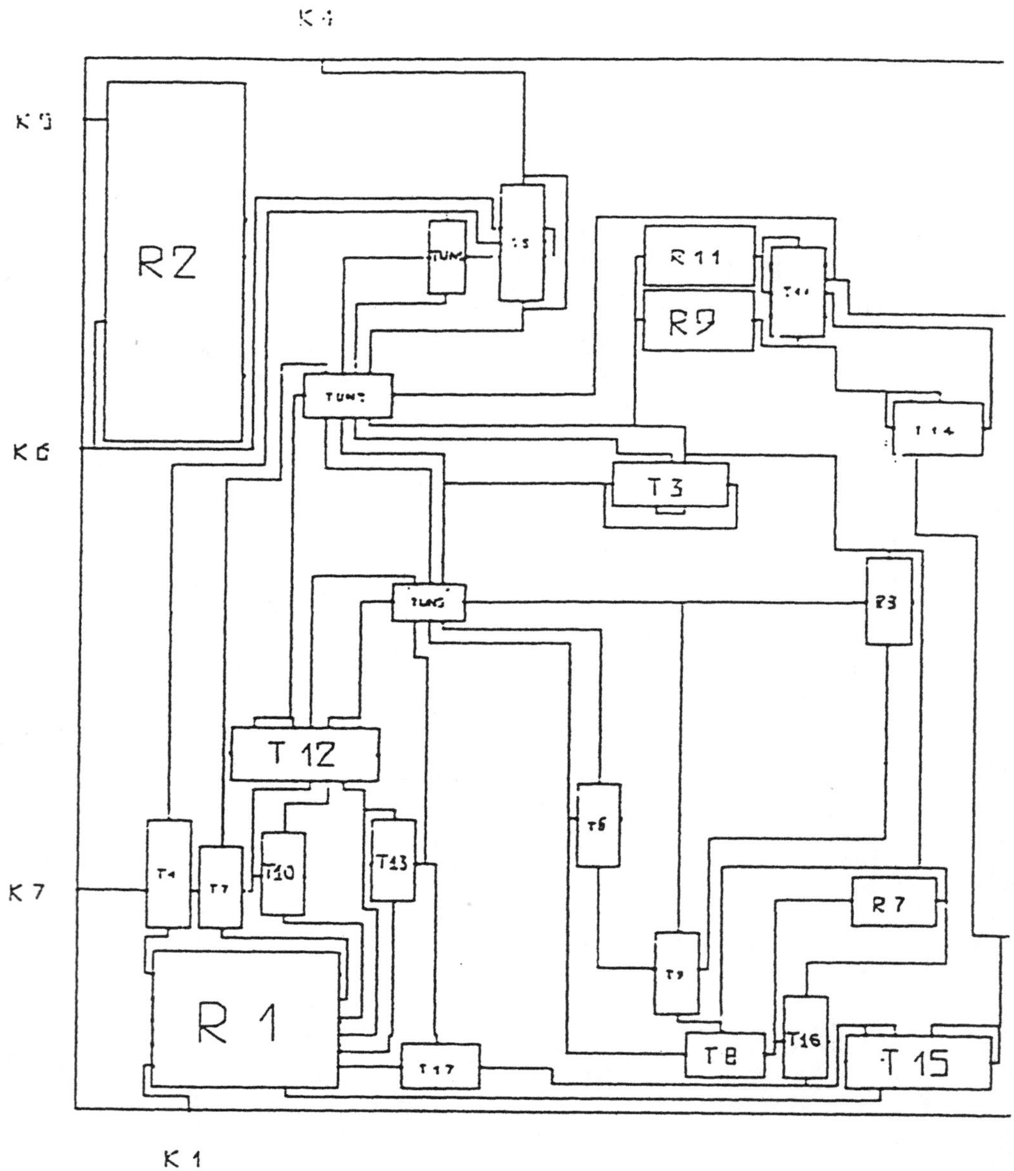

Fig. 7.24. The intermediate layout sketch obtained from that of Figure 7.23 The additional routing was implemented by diffused tunnel $TUN3$ placement. The layout was optimized by auto-interactive commands TURN, SHIFT, REROUTING.

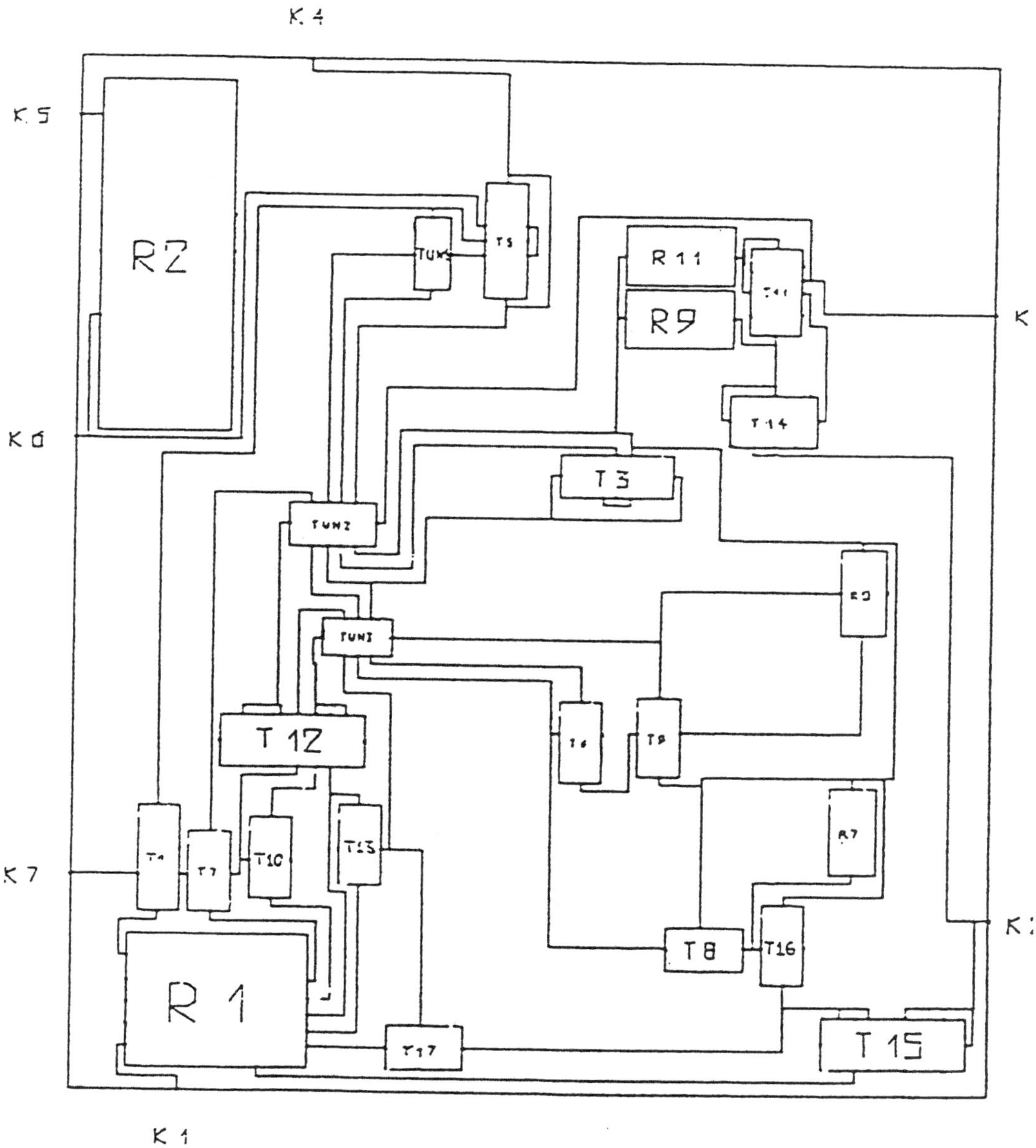

Fig. 7.25. The intermediate layout version obtained from that of Figure 7.24 Partial layout optimization was done by SHIFT command for elements $TUN2$, $TUN3$, $T8$, $T9$ and TURN command for element $R7$.

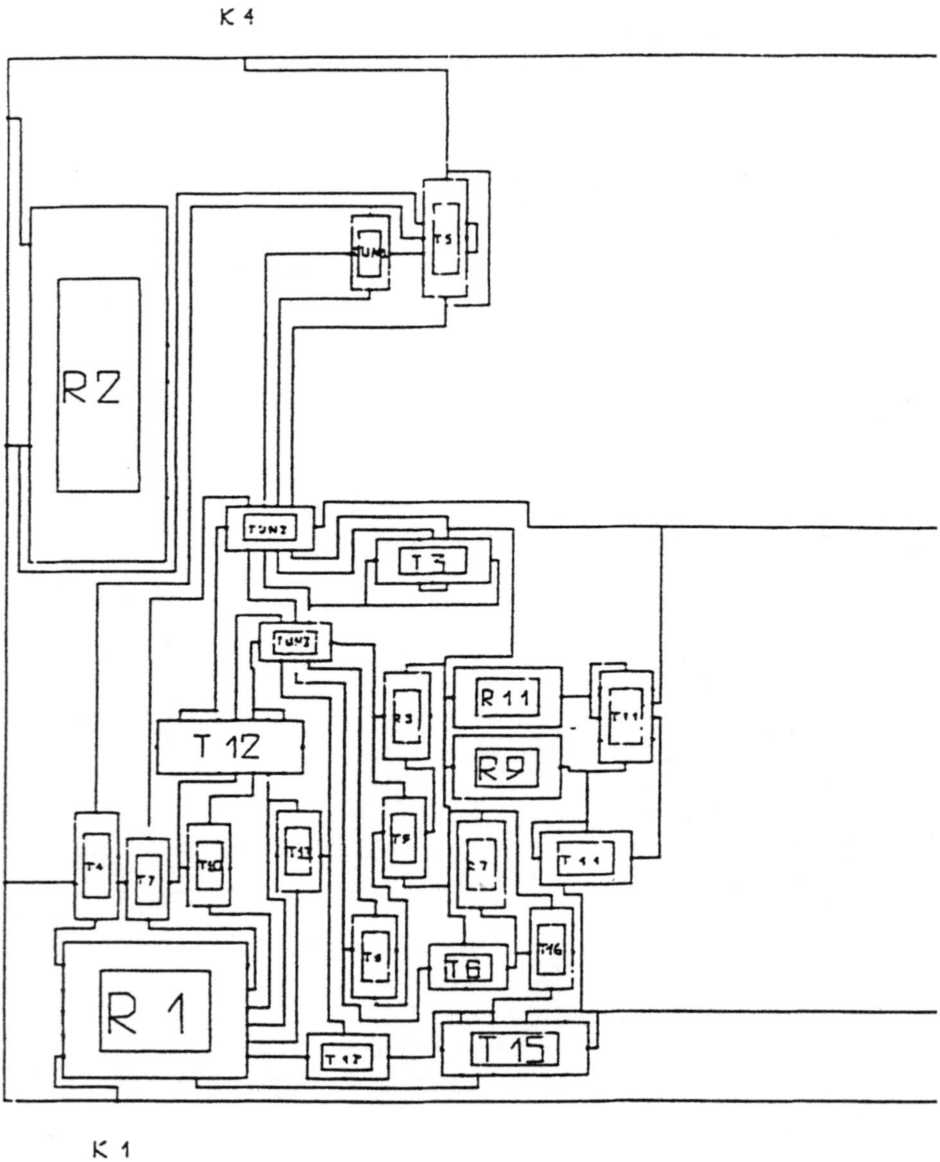

Fig. 7.26. The intermediate layout version obtained from that of Figure 7.25 Element fixing was done to keep their mutual placement during further actions. The elements with

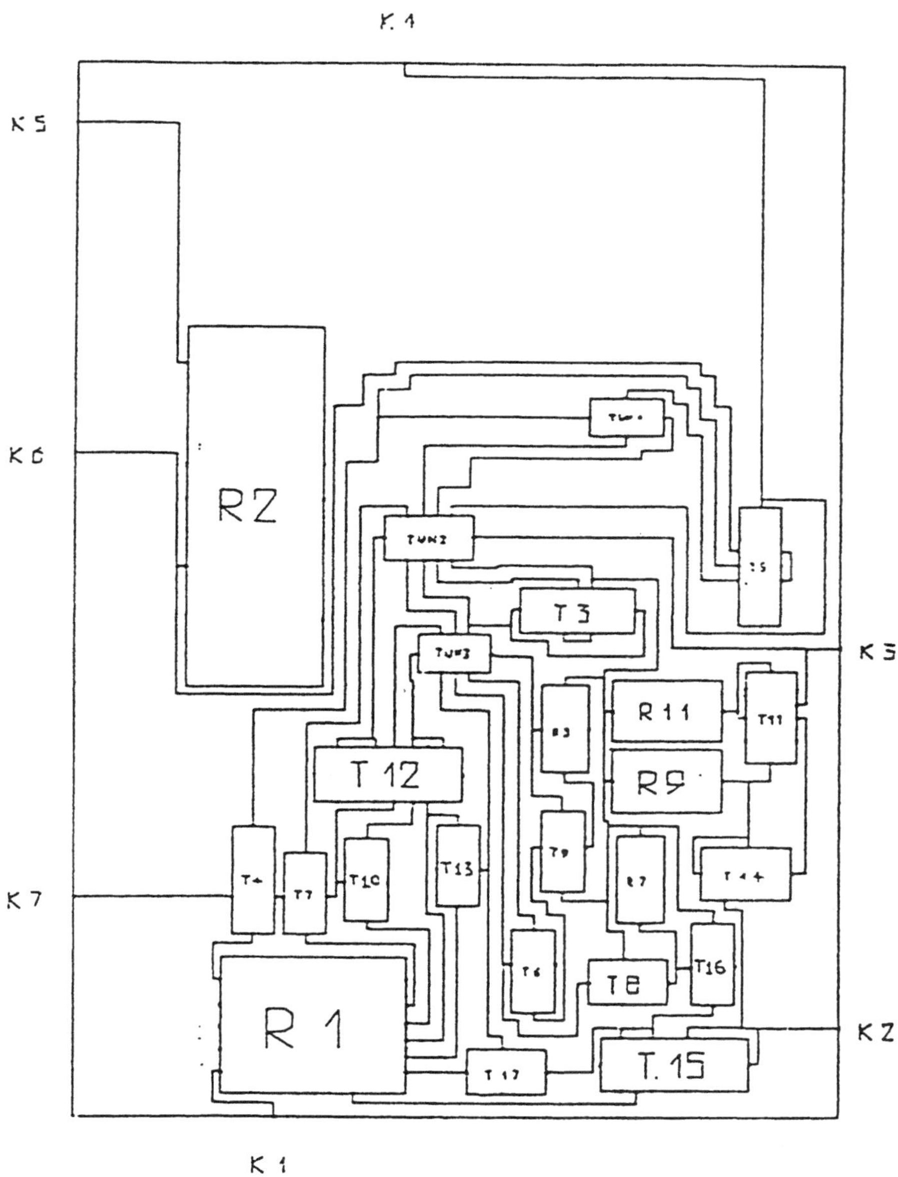

Fig. 7.27. The intermediate layout version obtained from that of Figure 7.26 by the SHIFT of the element $T5$ and $TUN1$.

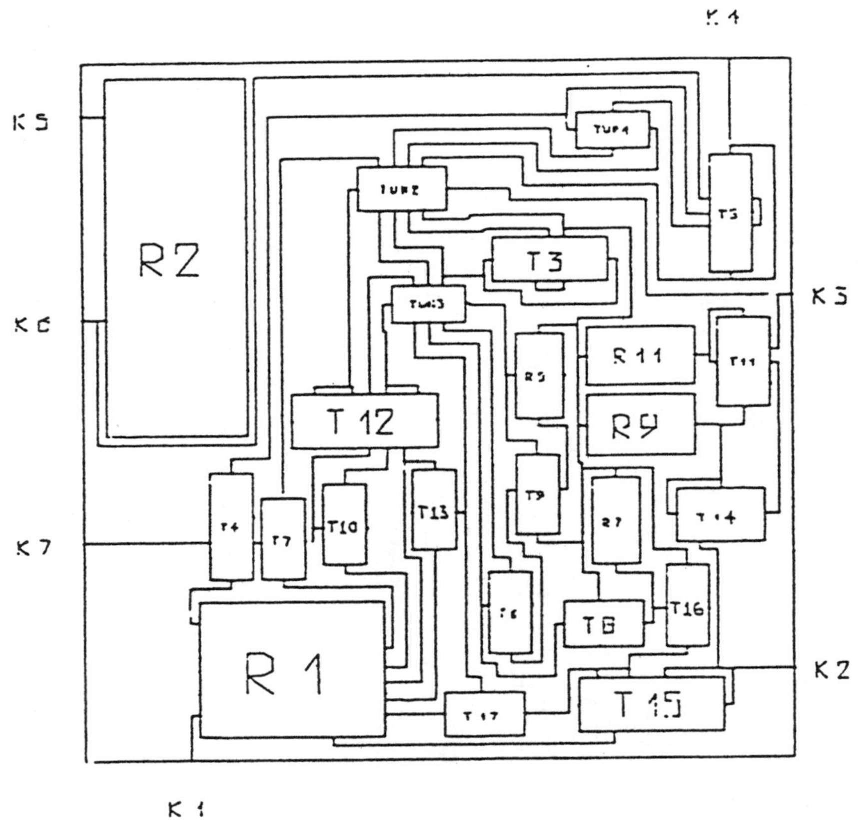

Fig. 7.28. The layout sketch was obtained from that of Figure 7.27 by COMPACTION command.

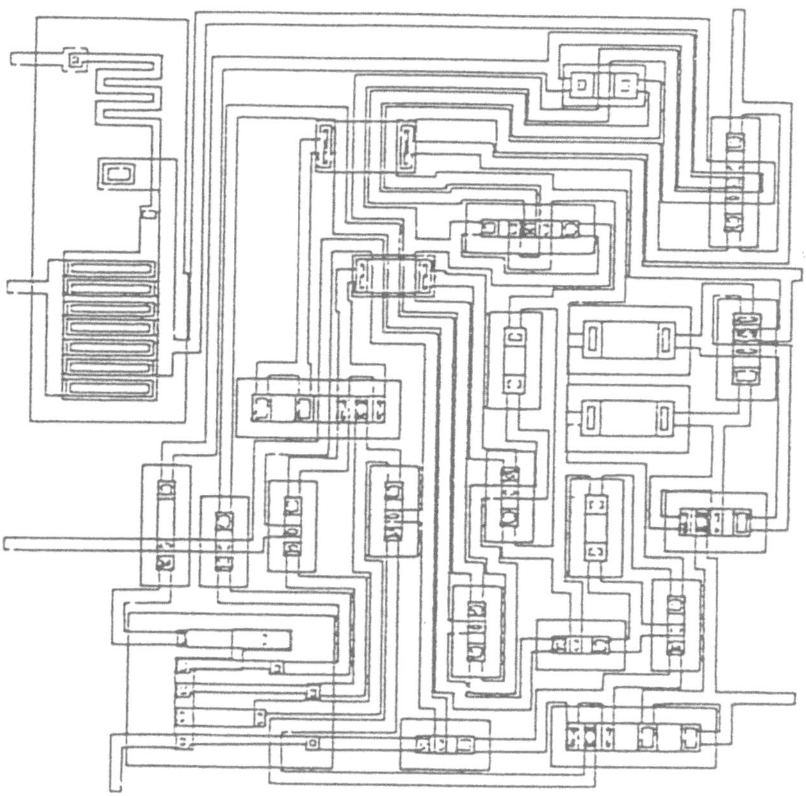

Fig. 7.29. The final version of the layout obtained in automated mode from the sketch of Figure 7.28 and library component.

References

1. AUSLANDER, L., PARTER, S.V., *On embedding graphs in the plane*, J. Math. and Mech. 10 (1961), No.3, pp. 517-523.
2. AZARYONOK, A.S., *A study of some combinatorial problems of VLSI layout design*, Dis... kand. fiz.-mat. nauk, Minsk, (1990), in Russian.
3. AZARYONOK, A.S., SARVANOV, V.I., *On the complexity of planar realization of hypergraphs*, Vesci AN BSSR, Ser. fiz.-mat. nauk, (1987), No. 4, pp. 10-12, in Russian.
4. AZARYONOK, A.S., SARVANOV, V.I., *On the planarity of hypergraphs with limited parameters*, DAN BSSR, 34 (1990), No. 9, pp. 781-784, in Russian.
5. BADER, W., *Das topologische problem des gedruckten shaltung and seine-losung*, Arch. Elekt., 49 (1964), No. 1, pp. 2-12.
6. BATALOV, B.V., KAZENKOV, G.G., KURMAJEV, F.A., SHCHEMILININ, V.M., *Algorithm for placement of IC component with the minimum number of internal connections*, Mikroelectronika, 3 (1969), pp. 282-302, in Russian.
7. BERGE, C., *Graphs and Hypergraphs*, Amsterdam, (1973).
8. BOOTH, K.S., LUEKER, G.S., *Testing for the consecutive ones property, interval graphs and graph planarity using PQ-tree algorithms*, J. Comp. Syst. Sci., 13 (1976), No. 3, pp. 335-379.
9. BRUNO, J., STEIGLITZ, K., WEINBERG, L., *A new planarity test based on 3-connectivity*, IEEE Trans., CT-17 (1970), No. 2, pp. 197-206.
10. CARSON, D.I., *On $O(p^2)$ algorithms for planarization*, IEEE Transactions on Computer-Aided Design, 12 (1993), No. 9, pp.1300-1302.
11. CHEN, H.-F.S., LEE, D.T., *A Faster Algorithm for Rubber-Band Equivalent Transformation for Planar VLSI Layouts*, IEEE Transactions on Computer-Aided Design, 15 (1996), No.2, pp.217-227.

12. Chen, T., Kang, T., *The connectivity of hypergraph and the design of fault-tolerant multibus systems*, 18Th. Int. Symp. Fault-tolerant Computing FTCS-18, Tokyo, June 27-30, (1988), pp. 374-379.
13. Chiba, T., Nishizeki, T., Ale, S., Ozawa, T., *A linear algorithm for embedding planar graphs using PQ-trees*, J. Comp. Syst. Sci., 30 (1985), No. 1, pp. 54-76.
14. Chiba, T., Nichioka, I., Shirakawa, I. *An algorithm of maximal planarization of graphs*, Proc. 1979 IEEE Int. Symp. on Circuits and Systems, (1979), pp. 649-652.
15. Cole, R., Siegel, A., *River routing every which way, but loose*, in Proc. 25th Annual Symp. Found. Comput. Sci., 1984, pp.65-73.
16. Cong, J., Hossain, M., Sherwani, N.A., *A provably good multilayer topological planar routing algorithm in IC layout designs*, IEEE Transactions on Computer-Aided Design, 12 (1993), No. 1, pp. 70-78.
17. Demucron, G., Malgrange, J., Pertuiset, R., *Graphs planaires reconnaissause et construction des representation planaires topologigues*, Rev. Francais de Rech. Operat., 8 (1964), pp. 34-37.
18. Dai, W.W.-M., Kong, R., Jue, J., *Rubber band routing and dynamic data representation*, in Proc. IEEE Int. Conf. Computer-Aided Design, Santa Clara, CA, 1990, pp.52-55.
19. Dai, W.W.-M., Kong, R., Sato, M., *Routability of a rubber-band sketch*, in Proc. 28th ACM/IEEE Design Automation Conf., 1991, pp. 45-48.
20. Dai, W.W.-M., Dayan, T., Staepelaere, D., *Topological routing in SURF: Generating a rubber-band sketch*, in Proc. 28th ACM/IEEE Design Automation Conf., 1991, pp. 39-44.
21. Dolev, D., Karplus, K., Siege, A., Strong, A., Ullman, J.D., *Optimal wiring between rectangle*, Proceedings of 13th Annual ACM Symposium on Theory of Computation, May 1987, pp. 312-317.
22. Engl, W.L., Mlynaki, D.A., Pernards., P., *Theory of Multiplace Graphs*, IEEE Trans. Circuits and Syst. (1975), No. 1.
23. Even, S., Tarjan, R.E., *Computing an st-numbering*, Theo. Comp. Sci., 2 (1976), pp. 339-344.
24. Feinberg, V.Z., *Auto-interactive design of LSI layout*, Doct. Dis., Minsk, (1988), in Russian.
25. Feinberg, V.Z., Landa, A.B., Rabinovich, E.B., *LSI layout design automation system*, Design automation in Radio Engineering and Computer Science, M., MDNTP, (1987), pp. 112-116, in Russian.
26. Feinberg, V.Z., Rabinovich, E.B., Svirnovsky, Y.D., *A problem*

of identifying the maximal planar subgraph in LSI circuit planarization, Proc. of All-Union Res. Dev. Conference, Kaunas, (1989), in Russian.

27. FEINBERG, V.Z., RABINOVICH, E.B., SVIRNOVSKY Y.D. (1989) *LSI circuit planarization system*, Elektron. tehn., Issue 6 "Mikroelekronika", (1989), in Russian.
28. FISCHER, G.J., WING, O., *Computer recognition and extraction of planar graph from the incidence matrix*, IEEE Trans., CT-13 (1966), pp. 154-163.
29. FLAMENT, C., *Hypergraphs Arbores*, Discrete Mathematics, 21 (1978), No. 3, pp. 223-227.
30. GAREY, M.R., JOHNSON, D.S., *Computers and Intractability: A Guide to the Theory of NP-Completeness*, W.H.Freeman and company, San Francisco, (1979).
31. GOLDSTEIN, A.J., SCHWEIKER, D.G., *A proper model for testing the planarity of electrical circuits*, The Bell System Technical J., 52 (1973), No. 1, pp. 135-142.
32. HARARY, F., *Graph Theory*, Addison-Wesley Publishing Company, (1969).
33. HARARY, F., TUTTE, W.T., *A Dual Form of Kuratowski's Theorem*, Canad. Math. Bull., 8 (1965), pp. 17-20,373.
34. HO, J.M., VIJAYAN, G., WONG, C.K., *Planar topological routing of pad nets*, Integration, 11 (1991), No. 3, pp.295-326.
35. HO, J.M., SARRAFZADEH, M., VIJAYAN, G., WONG, C.K., *Pad minimization for planar routing of multiple power nets*, IEEE Transactions on Computer-Aided Design, 9 (1990), No. 4, pp. 419-426.
36. HO, J.M., SUZUKI, A., SARRAFZADEH, M., *An exact algorithm for single-layer wire length minimization*, IEEE Transactions on Computer-Aided Design, 12 (1993), pp. 175-180.
37. HOPCROFT, J., TARJAN, R., *Efficient planarity testing*, J. Assoc. Comput. Machinery, 21 (1974), No. 4, pp. 549-569.
38. JAYAKUMAR, R., THULASIRAMAN, K., SWAMY, M.N.S., *Planar embedding: linear-time algorithms for vertex placement and edge ordering*, IEEE Trans. Circuits and Systems, CAS-35 (1988), No. 3, pp. 334-344.
39. JAYAKUMAR, R., THULASIRAMAN, K., SWAMY, M.N.S., $O(n^2)$ *algorithms for graph planarization*, IEEE Trans. Comput. Aid. Des. Integr. Circuits and Syst., 8 (1989), No.3, pp. 257-267.
40. JOHNSON, D.S., POLLAK, H.O., *Hypergraph Planarity and the Complexity of Drawing Venn Diagrams*, J. Graph Theory, 11 (1987), No. 3, pp. 309-325.

41. KARZANOV, A.V., *Linear-time planarity identification and embedding of a graph in the plane*, Discrete Optimization Algorithms and their Application in Computing Systems, Yaroslavl, (1983), pp. 58-80, in Russian.

42. KNUER, B., *A simple planarity criterion*, J. Assoc. Comp. Math., 22 (1975), No. 2, pp. 226-230.

43. KORSHUNOV, A.D., *Basic properties of random graphs with a great number of vertices and edges*, Uspehi mat. nauk, 40 (1985), Issue 1(241), pp. 107-173, in Russian.

44. KOU, L., MARKOWSKY, G., BERMAN, L., *A faster algorithm for the Steiner trees*, Acta Informatica, 15 (1981), pp. 141-145.

45. LEE, D.T., PREPARATA, F.P., *Euclidean shortest paths in the presence of rectilinear barriers*, Networks, 14 (1984), pp. 393-410.

46. LEISERSON, C.E., MALEY, F.M., *Algorithms for routing and testing routability of planar VLSI layouts*, in Proc. 17th Annual ACM Symp. Theory Computing, 1985, pp. 69-78.

47. LEISERSON, C.E., PINTER, R.Y., *Optimal Placement for river routing*, SIAM Journal of Computing, 12 (1983), No.3, pp. 447-462.

48. LEMPEL, A., EVEN, A., CEDERBAUM, I., *An algorithm for planarity testing of graphs*, in: Theory of Graphs, Intern. Symp. Rome. Ed. P.Rosenstishl, Pub. Gordon and Breach, (1966), pp. 215-232.

49. LEVIN, A.G., *On critical nonplanar hypergraphs*, Vesci AN BSSR, Ser. fiz.-mat.navuk, Deposited manuscript, VINITI No.5553-B87, Aug. 3, (1987), in Russian.

50. LEVIN, A.G., *Some problems of hypergraph planarity*, Preprint, ITK AN BSSR, (1988), No.50, in Russian.

51. LEVIN, A.G., MELNOKOV, O.I., *Relaxed Helly property and hypergraph planarity*, Vesci AN BSSR, Ser. fiz.-mat. navuk, (1988), No. 3, pp. 12-15, in Russian.

52. LEVIN, A.G., RABINOVICH, E.B., *Approximation algorithms of hupergraph planarization*, Vesci AN BSSR, Ser. fiz.-mat. navuk, (1992), No. 2, pp. 110-116, in Russian.

53. LIAO,K.-F., SARRAFZADEH, M., *Boundary single-layer routing with movable terminals*, IEEE Transactions on Computer-Aided Design, 10 (1991), No. 11, pp. 1382-1391.

54. MALEY, F.M., *Single-Layer Wire Routing and Compaction*, The MIT Press, 1990.

55. MAREK-ZADOWSKA, M., *Pad assignment for power nets in VLSI circuits*, IEEE Transactions on Computer-Aided Design, CAD-6 (1987), No. 4, pp. 550-560.

56. NISHIZEKI, T., *Planar graphs problems*, Computing, 43 (1990), No.7, pp. 53-68.
57. NISHIZEKI, T., CHIBA, N., *Planar Graphs: Theory and Algorithms*, Annals of Discrete Math., 32 (1988), 232 p.
58. ORE, O., *The four-color problem*, Academic Press, New York - London, (1967).
59. OZAWA, T., TAKAHASHI, H., *A graph planarization algorithm and its application to random graphs*, in Graph Theory and Algorithms, Springer-Verlag Lecture Notes in Computer Science, 108 (1981), pp. 95-107.
60. PAPADIMITRIOU, C.H., STEIGLITZ, K., *Combinatorial Optimization: Algorithms and Complexity*, Prentice-Hall Inc, Englewood Cliffs, New Jersey, (1982).
61. PETRENKO, A.I., TETELBAUM, A.Y., *Electronic device models in solving engineering problems*, Kibernetika, (1979), No. 2, pp. 47-54, in Russian.
62. PETRENKO, A.I., TETELBAUM, A.Y., SHRAMCHENKO, V.L., *Computer-aided design of electronic equipment (topological approach)*, Kijev, Vyshcha shkola, (1980), in Russian.
63. PILIPOSYAN, T.E., *Minimal hypergraph realizations and minimal placement of some graph classes*, Dis... kand. fiz.-mat. nauk, Minsk, (1983), in Russian.
64. PILIPOSYAN, T.E., *On planar- realized hypergraphs*, DAN ArmSSR, 82 (1986), No. 3, pp. 102-106, in Russian.
65. PRATUSJEVICH, M.L., *Improved mathematical models for constructing planar embedding of circuits*, Design Automation in Electronic Engineering, Kijev, Issue 24 (1981), pp. 15-22, in Russian.
66. PRATUSJEVICH, M.L., SELJUTIN, V.A., *Constructing planar circuit embeddings with account of design constraints*, Mikroelektronika, (1980), No. 3, pp. 202-208, in Russian.
67. RABINOVICH, E.B., FEINBERG, V.Z., *On planar LSI embedding*, Proc. of All-Union Res. Devel. Conference "Methods and Techniques of Noise Suppression in Digital Electronics", Kaunas, pp. 65-66, in Russian.
68. RICHARDS, D., *Complexity of single-layer routing*, IEEE Transactions on Computers, C-33 (1984), No.3, pp.286-288.
69. ROSKIND, J., TARJAN, R., *A Note on Finding Minimum-Cost Edge-Disjoint Spanning Trees*, Math. of Oper. Research, 10 (1985), No. 4, pp. 701-708.
70. RUBIN, F., *An improved algorithm for testing the planarity of a graph*,

IEEE Trans., 24 (1975), No. 2, pp. 113-121.

71. SELJUTIN, V.A., *Topological models for switching circuit design problems*, Elektron. tehnika, Ser.6, Issue 6 "Mikroelektronika", (1969), pp. 66-72.
72. SARRAFZADEH, M., LIAO, K.-F., WONG, C.K., *Single-layer global routing*, IEEE Transactions on Computer-Aided Design, 13 (1994), pp. 38-47.
73. SHRAMCHENKO, B.L., *The use of graph planarity criterion in computer-aided circuit design*, Kibernetika, (1989), No. 2, pp. 37-41, pp. in Russian.
74. SLATER, P., *A Characterization of SOFT Hypergraphs*, Canad. Math. Bull., 21 (1978), No. 3, pp. 335-337.
75. SUPOWIT, K.J., *Finding a maximum planar subset of a nets in a channel*, IEEE Transactions on Computer-Aided Design, 6 (1987), No. 1, pp. 93-94.
76. TAKAHASHI, H., MATSUYAMA, A., *An approximate solution for the Steiner problem in graphs*, Math. Japonica, 24 (1980), No. 6, pp. 573-577.
77. TETELBAUM, A.Y., SHRAMCHENKO, B.L., *Computer-aided design methods for electronic devices (survey)*, Zarub. radioel., (1977), No. 2, pp. 24-50, in Russian.
78. TETELBAUM, A.Y., SHRAMCHENKO, B.L., *Hypergraph application for planarity testing of circuits*, Izv. AN SSSR, Tehn. kib., (1978), No. 5, pp. 127-136, in Russian.
79. TOMPA, M., *An optimal solution to a wire-routing problem*, Journal of Computer and System Sciences, 23 (1981), No.2, pp.127-150.
80. VANLIER, M.C., OTTEN, R.H.J.M., *Automatic IC Layout: The Model and Technology*, IEEE Transaction on Circuits and Systems, CAS-23 (1976), No. 12, pp. 845-854.
81. VANCLEEMPUT, W.M., *Mathematical Models for the Circuit Layout Problem*, IEEE Transaction on Circuits and Systems, CAS-22 (1976), No. 11, pp. 759-767.
82. VOLODINA, V.V., CHERKASSKY, E.V., *On implementation of an algorithm for planarity testing and graph embedding in the plane for $O(v)$ operations*, Algorithm and Solutions for Network Problems, M., TSEMI, pp. 59-91, in Russian.
83. VOLOSHINA, A.A., FEINBERG, V.Z., *On hypergraph planarity*, DAN BSSR, 28 (1984), No. 4, pp. 309-311.
84. WAGNER, K., *Uber eine Eigenschaft der ebenen Komplexe*, Math. Ann., 114 (1937), pp. 570-590.

85. WEINSTEIN, Y.V., *An algorithm of complete planarization of circuit graphs*, Soob. AN GSSR, 122 (1986), No. 1, pp. 53-55, pp. in Russian.
86. WEINSTEIN, Y.V., *Considering orientation of component pins in the algorithm for circuit graph planarization*, Soob. AN GSSR, 127 (1987), No. 3, pp. 501-503, in Russian.
87. WILSON, R.J., *Introduction to Graph Theory*, Oliver and Boyd, Edinburgh, (1972).
88. WINTER, P., *Steiner problem in networks: a survey*, Networks, 17 (1987), 129-167.
89. WHITNEY, H., *Non-separable and planar graphs*, Trans. Amer. Math. Soc., 34 (1932), pp. 339-362.
90. WHITNEY, H., *A set of topological invariants for graphs*, Amer. J. Math., 55 (1933), pp. 231-235.
91. YEMELICHEV, V.A., MELNIKOV, O.I., SARVANOV, V.I, TYSHKEVICH, R.I., *Lectures on graph theory*, M., Nauka, (1990), in Russian.
92. ZYKOV, A.A., *Hypergraphs*, Uspehi mat. nauk, 22 (1974), No. 6, pp. 89-154, in Russian.

Index

Other *Mathematics and Its Applications* titles of interest:

W.L. Miranker: *Numerical Methods for Stiff Equations and Singular Perturbation Problems*. 1980, 220 pp. ISBN 90-277-1107-0

K. Rektorys: *The Method of Discretization in Time and Partial Differential Equations*. 1982, 470 pp. ISBN 90-277-1342-1

L. Ixary: *Numerical Methods and Differential Equations and Applications*. 1984, 360 pp. ISBN 90-277-1597-1

B.S. Razumikhin: *Physical Models and Equilibrium Methods in Programming and Economics*. 1984, 368 pp. ISBN 90-277-1644-7

A. Marciniak: *Numerical Solutions of the N-Body Problem*. 1985, 256 pp. ISBN 90-277-2058-4

Y. Cherruault: *Mathematical Modelling in Biomedicine*. 1986, 276 pp. ISBN 90-277-2149-1

C. Cuvelier, A. Segal and A.A. van Steenhoven: *Finite Element Methods and Navier-Stokes Equations*. 1986, 500 pp. ISBN 90-277-2148-3

A. Cuyt (ed.): *Nonlinear Numerical Methods and Rational Approximation*. 1988, 480 pp. ISBN 90-277-2669-8

L. Keviczky, M. Hilger and J. Kolostori: *Mathematics and Control Engineering of Grinding Technology. Ball Mill Grinding*. 1989, 188 pp. ISBN 0-7923-0051-3

N. Bakhvalov and G. Panasenko: *Homogenisation: Averaging Processes in Periodic Media. Mathematical Problems in the Mechanics of Composite Materials*. 1989, 404 pp. ISBN 0-7923-0049-1

R. Spigler (ed.): *Applied and Industrial Mathematics*. Venice-1, 1989. 1991, 388 pp. ISBN 0-7923-0521-3

C.A. Marinov and P. Neittaanmaki: *Mathematical Models in Electrical Circuits. Theory and Applications*. 1991, 160 pp. ISBN 0-7923-1155-8

Z. Zlatev: *Iterative Improvement of Direct Solutions of Large and Sparse Problems*. 1991, 328 pp. ISBN 0-7923-1154-X

M. Vajtersic: *Algorithms for Elliptic Problems*. 1992, 352 pp. ISBN 0-7923-1918-4

V. Kolmanovskii and A. Myshkis: *Applied Theory of Functional Differential Equations*. 1992, 223 pp. ISBN 0-7923-2013-1

A.D. Egorov, P.I. Sobolevsky and L.A. Yanovich: *Functional Integrals: Approximate Evaluation and Applications*. 1993, 426 pp. ISBN 0-7923-2193-6

G.I. Marchuk: *Adjoint Equations and Analysis of Complex Systems*. 1994 ISBN 0-7923-3013-7

Other *Mathematics and Its Applications* titles of interest:

A. Bakushinsky and A. Goncharsky: *Ill-Posed Problems: Theory and Applications.* 1994, 256 pp. ISBN 0-7923-3073-0

G.N. Milstein: *Numerical Integration of Stochastic Differential Equations.* 1995, 170 pp. ISBN 0-7923-3213-X

V.V. Shaidurov: *Multigrid Methods for Finite Elements.* 1995, 331 pp. ISBN 0-7923-3290-3

A.N. Tikhonov, A.V. Goncharsky, V.V. Stepanov and A.G. Yagola: *Numerical Methods for the Solution of Ill-Posed Problems.* 1995, 264 pp. ISBN 0-7923-3583-X

H.W. Engl, M. Hanke and A. Neubauer: *Regularization of Inverse Problems.* 1996, 328 pp. ISBN 0-7923-4157-0

G.I. Marchuk: *Mathematical Modelling of Immune Response in Infectious Diseases.* 1997, 358 pp. ISBN 0-7923-4528-2

V. Feinberg, A. Levin and E. Rabinovich: *VLSI Planarization.* 1997, 190 pp. ISBN 0-7923-4510-X

MIX

Papier | Fördert gute Waldnutzung

FSC® C083411

Kolibri 360 GmbH
Ferdinand-Jühlke-Straße 7
99095 Erfurt, Deutschland
produktsicherheit@kolibri360.de